TREE HARVESTING TECHNIQUES

FORESTRY SCIENCES

Baas P, ed: New Perspectives in Wood Anatomy. 1982. ISBN 90-247-2526-7

Prins CFL, ed: Production, Marketing and Use of Finger-Jointed Sawnwood. 1982. ISBN 90-247-2569-0

Oldeman RAA, et al., eds: Tropical Hardwood Utilization: Practice and Prospects. 1982. ISBN 90-247-2581-X

Den Ouden P and Boom BK, eds: Manual of Cultivated Conifers: Hardy in Cold and Warm-Temperate Zone. 1982. ISBN 90-247-2148-2 paperback; ISBN 90-247-2644-1 hardbound.

Bonga JM and Durzan DJ, eds: Tissue Culture in Forestry. 1982. ISBN 90-247-2660-3

Satoo T and Magwick HAI: Forest Biomass. 1982. ISBN 90-247-2710-3

Van Nao T, ed: Forest Fire Prevention and Control. 1982. ISBN 90-247-3050-3

Douglas J, ed: A Re-appraisal of Forestry Development in Developing Countries. 1983. ISBN 90-247-2830-4

Gordon JC and Wheeler CT, eds: Biological Nitrogen Fixation in Forest Ecosystems: Foundations and Applications. 1983. ISBN 90-247-2849-5

Hummel FC, ed: Forest Policy: A Contribution to Resource Development. 1984. ISBN 90-247-2883-5

Duryea ML and Landis TD, eds: Forest Nursery Manual: Production of Bareroot Seedlings. 1984. ISBN 90-247-2913-0

Manion PD, ed: Scleroderris Canker of Conifers. 1984. ISBN 90-247-2912-2

Tree harvesting techniques

by

K.A.G. STAAF
College of Forestry
Swedish University of Agriculture
Uppsala, Sweden

and

N.A. WIKSTEN
Canadian Executive Overseas
Montreal, PQ, Canada

1984 **MARTINUS NIJHOFF/DR W. JUNK PUBLISHERS**
a member of the KLUWER ACADEMIC PUBLISHERS GROUP
DORDRECHT / BOSTON / LANCASTER

Distributors

for the United States and Canada: Kluwer Academic Publishers, 190 Old Derby Street, Hingham, MA 02043, USA
for the UK and Ireland: Kluwer Academic Publishers, MTP Press Limited, Falcon House, Queen Square, Lancaster LA1 1RN, England
for all other countries: Kluwer Academic Publishers Group, Distribution Center, P.O. Box 322, 3300 AH Dordrecht, The Netherlands

Library of Congress Cataloging in Publication Data

Staaf, K. A. G. (K. Anders G.)
Tree harvesting techniques.

(Forestry sciences)
A condensed edition of the original Swedish version of 1972.
Bibliography: p.
1. Logging. I. Wiksten, N. A. (N. Åke) II. Title.
III. Series.
SD538.S82155 1984 634.9'82 84-14692
ISBN 90-247-2994-7

ISBN 90-247-2994-7 (this volume)

PRINTED IN THE NETHERLANDS

Contents

Preface

The introduction of chain saws and tractors in the early 1950's marked the beginning of a change in tree harvesting techniques from the old manual methods to mechanized operations. It was followed by a rapid evolution both technically and systematically. Hence, the requirements for improved knowledge of operational efficiency also increased. Changing relations between Man, machines and environment brought about new experiences and awareness of a physiological and ergonomic nature. Improved knowledge of both machine technology and planning of work on a small or large scale has grown increasingly important for an efficient utilization of expensive machines and other equipment.

The need for a textbook on tree harvesting techniques including experiences made in recent years is enhanced. The book presented here is primarily based on lectures given on the subject of Forest Techniques at the Faculty of Forestry at the Swedish University of Agricultural Sciences and after modifications also at the University of Nairobi (Kenya). Thus, the book is written primarily for students at the faculties and institutes of forestry. However, it is also useful for persons actively occupied in forest operations.

The presentation of this book in its original Swedish version in 1972 created a considerable interest in the preparation of a condensed edition in English. Thus interest has been expressed in Finland, Norway, Holland, Canada, U.S.A., Brazil, Japan, Poland, Scotland and Yugoslavia.

The authors have interpreted the widespread interest in the previous edition as an interest in the techniques of tree harvesting as applied primarily in the Nordic countries.

Most references to various sources of basic information in the Swedish edition of 1972 were of Swedish, Norwegian, Danish and Finnish origin and written in the native languages only. A large part of it was based on lectures given at the Faculty of Forestry at the Swedish University of Agriculture. Although most of this background information is available in the native languages only, it has been considered desirable to quote and date the references in the English edition in case they are of interest for translation.

Working in close contact with specialists in several countries the authors have solicited their viewpoints on this attempt at preparing an international textbook on "Tree Harvesting Techniques". Naturally, it has been difficult to accommodate all wishes and recommendations without contradictions but we have tried to arrive at a certain balance of opinions on the basis of reason and logic. Still, in view of the common background of the authors, it can be reasonably expected that some of the statements have been subconsciously tainted by the conditions prevailing in the Nordic countries.

Although the title of the book has been made more specific than that of the first edition (in Swedish only), the authors have decided, space permitting, to retain an abbreviated review of the planning process in order to give a logical background to the choice of harvesting systems. Tree harvesting techniques is not only a matter of machines but also of the appropriate ways of operating the machines under various environmental and socio-economic conditions. The application of tree harvesting techniques depends on recommendations given in the plans for regulated harvest of forest products. A mere description of machines and their use would make the book just a catalogue of equipment. To write a book for specialists would be presumptuous, the specialists having a tendency to contradict each other depending on the special conditions of their own countries. There is no final word in science and there is no single machine, method or system that is correct for all situations and conditions. What may be right for the exploitation of old, big timber today over large areas will not be the right techniques for the harvest of timber of various assortments from new, more uniform stands of man-made forests.

The final sections of this edition in English have been devoted to a brief description of ergonomics and its importance for the rationalization of the tree harvesting operations.

For their comments and recommendations we are particularly grateful to the following persons: Professor Dr. Marten Bol, The Netherlands, Professor Dr. Branko Mihac, Yugoslavia, Professor Dr. Kalle Putkisto, Finland, Professor Dr. Ivar Samset, Norway and Dr. Hon. Ross Silversides, Canada.

A quick linguistic evaluation has been provided through the cooperation of the Editor of the Petawawa National Forestry Institute, Mr. Yapo, Canada.

To all these persons we want to express our sincere gratitude.

Technical illustrations have been reprocessed at the Faculty of Forestry mainly through the valuable and knowledgeable efforts by photographer Jonas Palm and artist/forest technician Sigurd Falk. Institute secretary Sigbritt Israelson has worked very energetically and persistently with typing, corrections and lay-out of the manuscript. The authors wish to express their most heartfelt gratitude to these three devoted co-workers at the Faculty.

Last, but not least, we wish to thank the Faculty of Forestry at the Swedish University of Agricultural Sciences for its support and assistance in various material ways.

Garpenberg, Hedemora
March, 1984

Anders Staaf

Åke Wiksten

Introduction

Trees are useful for a growing number of purposes. All the components of the trees from the outermost tips of the rootlets to the last needles or leaves on the twigs have a potential utility value. In addition to the conventional and dominating usefulness of trees as material for buildings and paper products, the wood, bark and chemicals in the trees have recently gained an increasing importance as partial replacements for our dwindling resources of easily accessible, non-renewable and expensive petro-chemical products. Fortunately, our forests with their trees can be renewed for improved and sustained yield by proper management.

However, trees are also beautiful. They may decorate our homes and gardens and embellish the shore lines, river banks and horizons. They cover large areas of the earth with green carpets of forests. The trees provide shelter and shade for our mammals, nesting places for our birds and fresh water for our fish. They stabilize the soils and the supply of clean water. The aesthetic and intangible values of some trees may sometimes be considered higher than their monetary value and such trees, therefore, should be saved from harvesting by a sensitive demarcation of the operations along natural boundaries. The need for preservation of landscape beauty must not be forgotten in our quest for improved living conditions and higher efficiency of timber harvesting operations.

Depending on the purpose of our forest production and the particular circumstances in each location, the harvesting of trees must apply various techniques, methods and systems developed and designed for complete utilization of the timber resources and for highest possible efficiency in the use of Man, Machines and Money. It is only through viable operations that continuity of high production can be sustained.

Tree harvesting – General

It seems logical at this stage to begin the textbook with some brief descriptions of basic terms and forms of production. We can then proceed without misunderstandings into planning of the forest operations using information on resources available, environmental conditions and characteristics of trees and forests that influence directly the performance of tree harvesting equipment.

Terminology

Tree harvesting is a technical term used in forestry to include all the partial operations from felling to transport of timber.

Tree harvesting techniques is a term encompassing the forms and techniques used in the partial operations. The word techniques, derived from the Greek word téchne, stands for the science of correct execution of trade, arts or the practical way of carrying out work. It may also include a set of rules, particular modes of operation, skills, or manual and mechanized methods applied in a work process.

Knowledge of tree harvesting techniques is part of the science of forest work.

The partial operations involved in tree harvesting are: felling, delimbing, bucking (cross-cutting), debarking, bunching and chipping.

Forwarding is a term used in tree harvesting for work involved in transporting or moving timber to landings usually located along permanent routes of transport.

Further transport. In this book the subject of tree harvesting has been given a slightly broader meaning. In addition to the partial operations and forwarding of timber from the stump to the landings, matters concerning further transport of timber have also been treated. Further transport is a transport from landings along roads or rivers to other transport routes or to places of marketing or industrial, centralized processing.

Forms of production

Dealing with a valuable crop of timber consisting of various assortments (sawlogs, standard length pulpwood, poles, chips etc), tree harvesting is an important form of production in forestry.

The production of timber in the forests is usually divided into primary and secondary production.

Primary production concerns the wood producing biological processes and silvicultural measures e.g. the establishment and treatments of stands designed to produce the highest possible volume and/or value of yield. This form of production is largely included in the concept of forest improvements.

Secondary production of timber involves the partial operations from felling to transport. Knowledge of this form of production may be included in the concept of tree harvesting.

Tree harvesting is initiated by planning on the basis of a large number of organizational considerations. Planning must also take into account the forest environment and its direct influence on various work operations.

A third form of production, which is occasionally called **tertiary production,** includes the activities involved in the conversion of timber at the industries.

The production of timber in the forest operations and the conversion of timber in the forest industries together constitute forestry. (Figure 1).

The various forms of production must be adjusted to one another, to the economic fluctuations and to other changes in the management conditions within forestry.

FIGURE 1. Production and conversion of timber together constitute forestry.

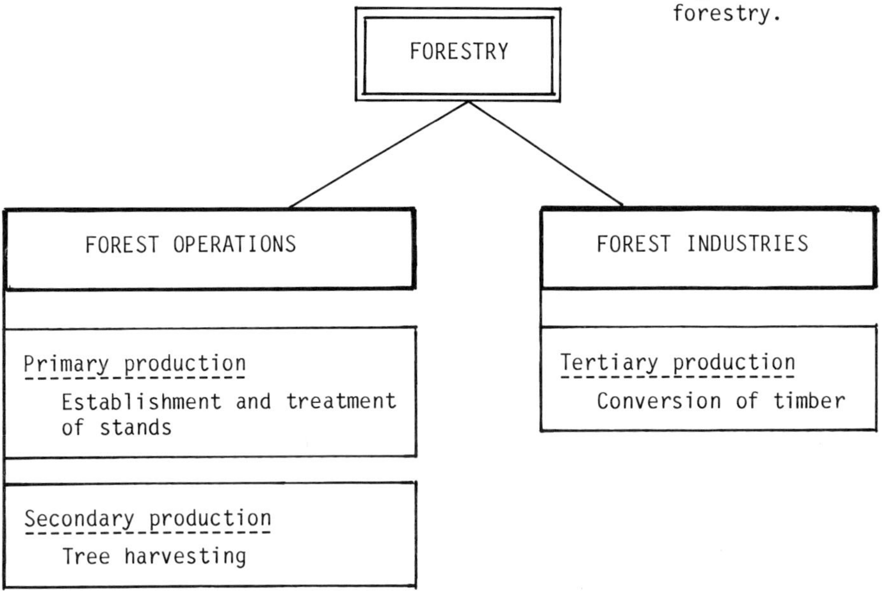

Rationalization of the various forms of operation

In recent years the various forms of operations in the forests have been subject to extensive rationalization because of adverse cost developments. This rationalization has been achieved due to conscientious, systematically designed programs aimed at improving the results of all forest activities.

Activities in the rationalization process have included mechanization with special emphasis on tree harvesting techniques. The economic development and a rapid technical advancement in general as well as growing demand for wood products and intensified competition encountered on the world market promoted the evolution of highly mechanized forest operations i.e. forest activities predominantly carried out by means of machines.

General objective

From a socio-economic point of view the general objective in the forests must be continuity of production as long as its importance to society is beyond doubt. The yield of timber, therefore, should be as stable as possible and it should be tied to the calculated requirements of the forest industries now and in the future.

Subordinate to the general objective, the prime goal of the tree harvesting operations can be defined as the achievement of lowest possible costs.

Tree harvesting a secondary form of production

The annual secondary production of timber by tree harvesting in our forests will amount to a certain regulated average volume ("allowable cut") that will ensure a sustained supply of timber. This volume varies slightly from one year to another depending on the composition of the harvest, market conditions, weather, accessibility of timber, availability of labour and suitable equipment etc.

Timber resources available. Depending on the state of resource development, the occurrence and age of stands established by conscientious methods of management (afforestation, reforestation, natural regeneration, thinning, selection systems etc), the annual secondary production will consist of various proportions of timber from exploitation of old forests and timber from managed forests. While timber from exploitation is characterized by a great variation in sizes, quality and soundness, timber from regularly managed forests is rather uniform in size and quality. This variation will influence the choice of tree harvesting methods and equipment.

Thinning and final harvest

In secondary production of timber in managed forests there are primarily two forms of harvest viz. *thinning* (young, small trees) and *final harvest* (old, large trees). In the exploitation of old forests final harvest operations dominate the secondary production.

Planning of tree harvesting

Objectives and means of planning

Planning requirements

Planning and control is required in order to achieve a good economic result of tree harvesting operations. Planning is a number of decisions concerning future activites. It leads to a program of actions that vary in nature and extent depending on the length of the plan and the size of the area concerned, forest activities required and the necessary resources of i.a. labour, machines, routes of transport etc.

In forest operations distinction is made between plans that are long term, intermediate term or short term. The intermediate term plans usually constitute the framework. General operation plans are designed in the major enterprises and forest management plans in the minor enterprises. Normally, the intermediate term plans encompass a period of ten years and they include partial plans concerning forest improvement work, tree harvesting, road construction, labour dispatch etc.

The importance of planning has increased in step with the production processes which the plan, in this case the tree harvesting plan, is supposed to govern.

Data required

Planning requires a total grasp of the current situation in the forests. Complete information is necessary in order to produce a tree harvesting plan which is mostly of a short term nature. Thus, data are required on i.a. the forest stands, terrain conditions, roads, population, labour supply, machines etc. Sample data concerning the forest stands and the sites can be collected for each stand by recording the area and the site quality as well as the age and volume of the trees.

Collection of stand data

Area of each stand is usually measured on air-photos. The volume of each stand is then obtained as a product of estimated volume of timber per hectare or acre and the area of the stand.

Estimates of various kinds are often based on measurements of trees randomly selected within sample plots or within inventory strips giving a required accuracy of about three percent. Information on roads and population distribution is obtained from public and socio-economic maps.

The collection of stand data etc. is followed by a calculation of the tree harvesting operations which is an estimate of the extent and composition of the activites for a given period of time and conditions i.a. concerning the principles of silvicultural measures.

Requirements of labour and machines

When the work load, i.e. primarily the amount of timber to be felled according to the tree harvesting plan, is estimated, the requirements of labour and machines are analyzed. This can be done primarily on the basis of known or calculated performance data, which in forest operations usually are expressed in number of man-days or machine-hours per m^3 of timber produced.

Preparation of the tree harvesting plan

After all the necessary data have been collected, a tree harvesting plan can be prepared for the achievement of a definite objective, usually in principle of an economic nature. Certainly, the objective of planning tree harvesting operations should be formulated before the collection of all facts is initiated.

Planning of tree harvesting operations is increasingly integrated (coordinated) with other activities associated with tree harvest such as storage and further transport of timber.

Desirability of short storage in the forests or at the industries and increased mechanization with a strong emphasis on full utilization of expensive machine equipment are facts pressing for improved tree harvesting plans. This improvement can now be achieved by means of mathematical models for the choice of tree harvesting systems (simulation) and by data processing.

Planning for low costs of tree harvesting

Generally, the main objective of planning is to achieve low costs of tree harvesting for a given annual operation. In this context it is possible to distinguish various conditions that promote low costs of tree harvesting e.g.

1. Full utilization of the machine equipment
2. Low costs in the various areas of tree harvesting by choice of proper season
3. Concentration of the tree harvesting operations in time and space

Maps

Planning and inventories in the forests are based on maps and air-photos of various kinds. The maps may be public maps, land survey maps, geological maps, meteorological maps and forest maps.

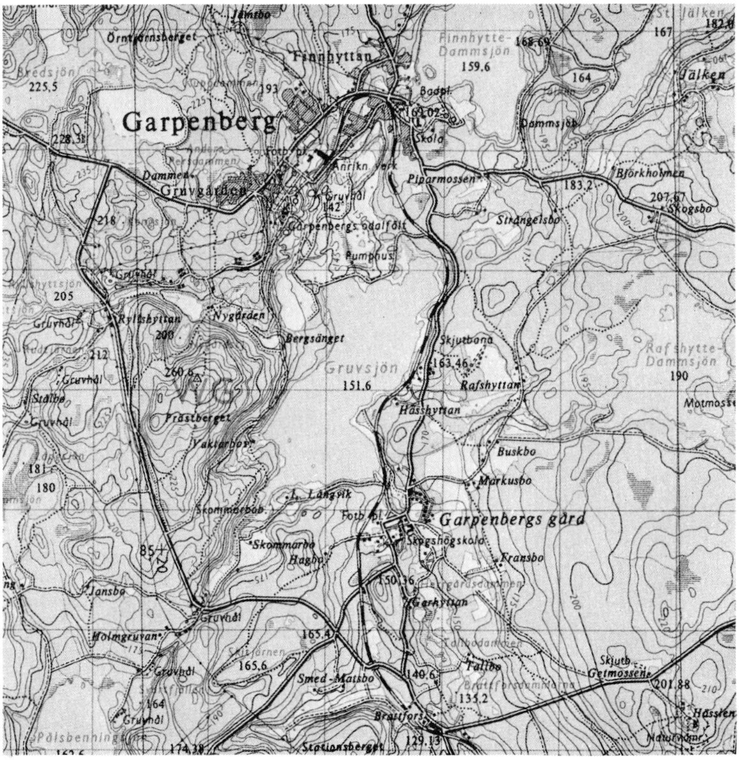

FIGURE 2. Separate print from a topographic map reproduced by a National map printshop.

Forest maps. Common forest maps are used i.a. as basic information at sampling and description of forest stands and for project proposals. Their use varies with the size of the forest, intensity of operations, size of management staff etc.

Within a given management unit the following requirements may be met by the use of forest maps:

- Overview of a certain area
- Guidance within the area
- Planning of measures in combination with notes and remarks of value for the activities within the area

New maps based on public maps and supplemented with a coordinate system are increasingly used for accurate information on e.g. the locations of storage places along the roads.

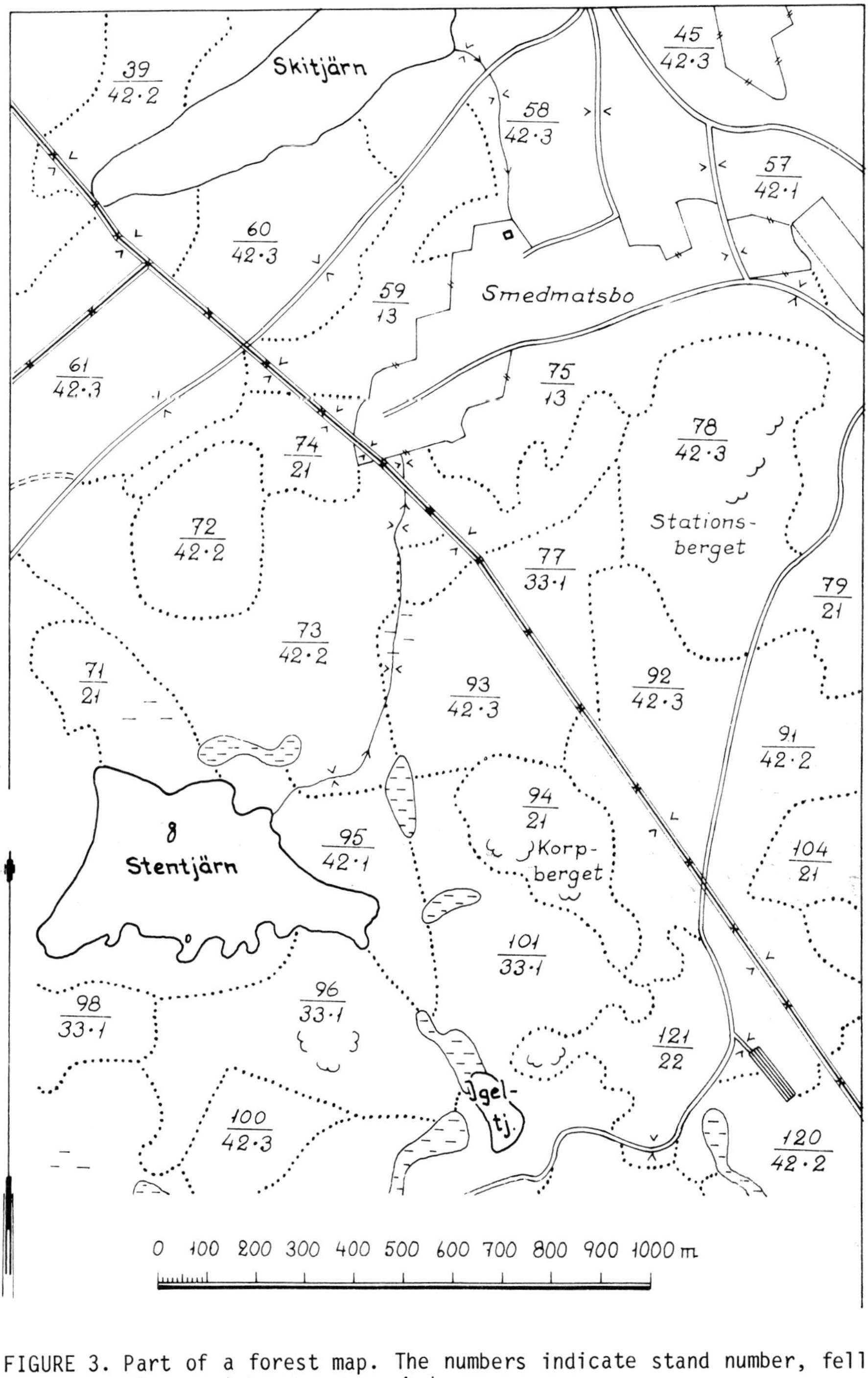

FIGURE 3. Part of a forest map. The numbers indicate stand number, felling class and treatment period.

Planning in general - Land and Labour

From the point of tree harvesting the structure and fragmentation of the forest properties i.e. their locations, shapes and sizes, are of great interest. The transition from manual work to mechanized forms of operation makes shape and areal distribution of the forest properties of great importance. Thus, the number of management units of private forest properties in a northern country in 1964 exceeded 260,000 of which approximately 64,000 were pure forest properties or properties where the owner had leased the farm land, keeping the forest for himself. The average area of forest land in the properties was approximately 11,500,000 : 260,000 = 44 hectares. In 1970 the corresponding average was 77 hectares. In conjunction with the current structural change in forest and farm operations a large number of management units are exchanged or consolidated.

The fragmentation of the properties varies depending on the history of property formation (Figure 4).

FIGURE 4. Overview map showing forest properties. Small narrow parcels make the application of mechanized tree harvesting operations difficult.

Various forms of cooperation

To reduce the disadvantages of extreme fragmentation of land and to achieve a better economic result of the operations, increased cooperation between the owners is now in effect. This cooperation is applied particularly in the private forests.

Cooperation can be organized in various ways i.a. by the creation of:

- Community forests
- Forest management areas
- Cooperation areas

Population and labour

Matters of population distribution. As a result of the current evolution from farming to industry and a service oriented society, the settlement of people has changed gradually from a scattered distribution to a concentration of people to larger communities and densely populated areas (urbanization). This change also affects the organization and daily activities in tree harvesting operations.

A concentration of the population, which means that the forest labour resides in large communities, results i.a. in longer travelling distances to and from the place of work.

Labour conditions. Forest work is carried ut primarily by two categories of labour. One category is composed of persons occupied in farming but working in the forests seasonally. The other category consists of persons who work in the forests almost year-round. To the first category of rather seasonal labour belong a large number of private forest owners. The group of year-round labour consists increasingly of permanently employed workers.

The changing proportions of total labour occupied with felling and processing of timber in large scale forestry in a northern country in the years of 1960, 1965 and 1969 are exemplified in Figure 5. Situation in February 1960 is given the relative number 100. (Bendz, Yttermyr, 1966).

Figure 5 also shows that seasonality of tree harvesting decreased during the period.

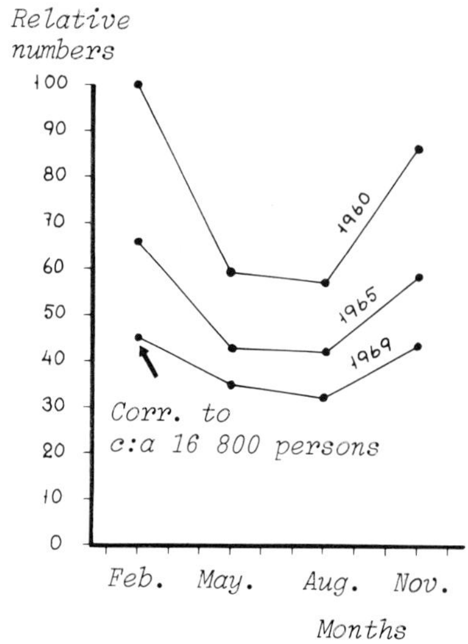

FIGURE 5. Labour occupied with tree harvesting in large-scale operations in 1960, 1965 and 1969 (Sweden).

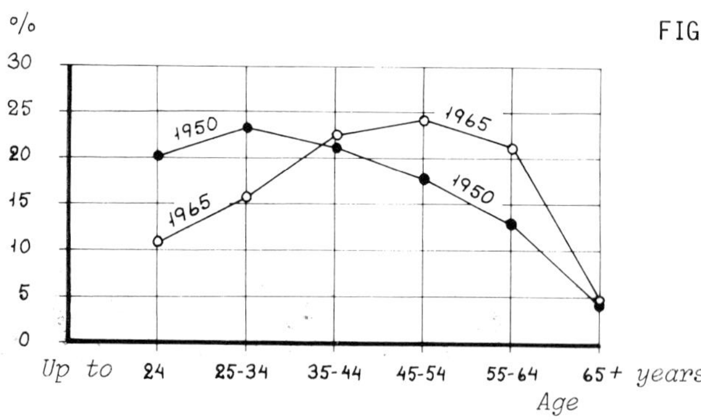

FIGURE 6. Example showing the age class distribution of the forest labour (Sweden).

Figure 6 shows how the age class distribution of the forest labour changed. How can this be explained? It could be the result of reduced recruitment of new labour to forest operations while the old labour has limited opportunities in other occupations and stays in forestry. Continued mechanization and rationalization will also reduce the total number of labour although a potential increase in the volume of timber harvested every year brought about by increased forest improvement work may have a compensating effect.

Areas and centres of labour

A major unit of operation is here presented in order to show the relationship between areas and centres of labour supply (Figure 7). The outline may represent a forest management unit with six forest guard (ranger) districts or a forest management area (private) with six districts (village units). (Staaf, 1960).

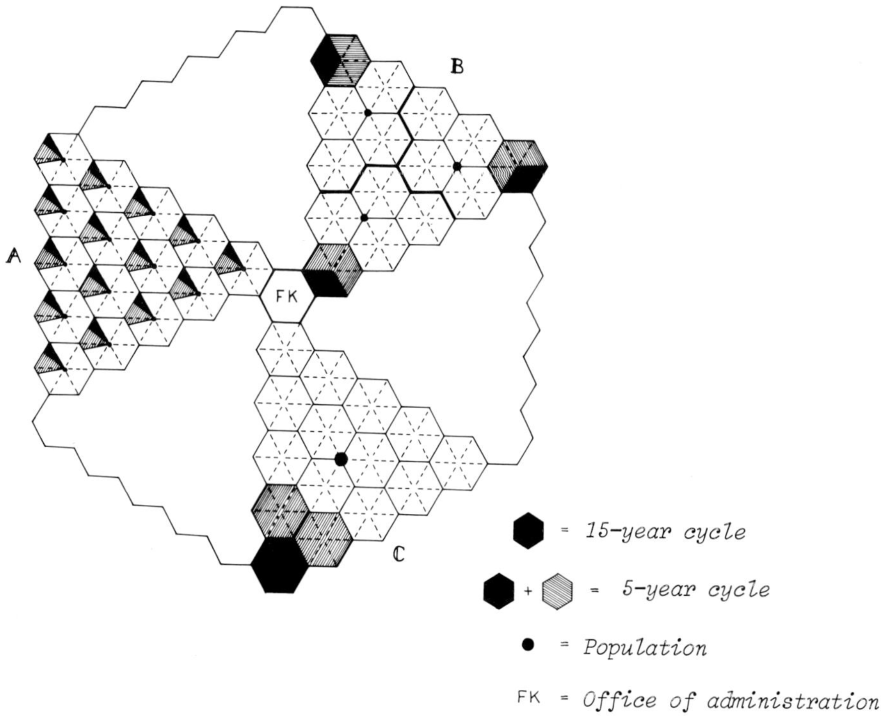

FIGURE 7. Outline of a major unit of operation.

Forest guard (ranger) district A

In forest guard (ranger) district A there are 15 small villages or scattered farms constituting centres of labour. Each centre has been assigned an area sufficient for the labour available in the centre. Each labour area is planned to contain a number of annual areas. It is delineated by natural boundaries as an area within which necessary measures such as tree harvesting, reforestation and forest improvement will be carried out in a given year.

One annual treatment area per village is the optimum area planned on the basis of the labour supply available in district A and its population distribution. Two or more annual areas per village or one annual area for two or more villages would cause losses from inoptimum situations, in the first case because of high costs of roads and supervision, in the latter case because of high costs of personnel transports.

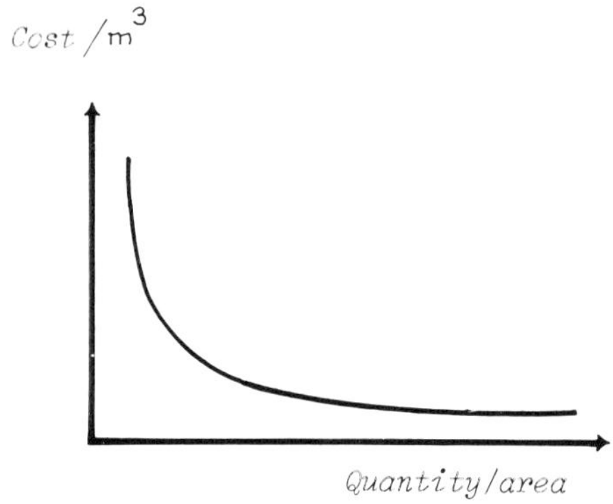

FIGURE 8. Influence of concentration on the costs of roads, supervision etc.

Concentration of tree harvesting work will reduce the costs of road maintenance. This will also reduce the costs of scaling and supervision per unit of timber volume as a result of reduced walking and travel time for the supervisory personnel. Figure 8 shows the general relationship between costs and concentration of work.

However, an increased concentration will raise the costs of transports of personnel and camp accomodations because of longer travel distances. These costs are higher when the road distance per hectare is short since the proportion of walking time is higher in relation to the travel time. The same applies to a scattered distribution of forest properties. An extremely scattered distribution of properties will mean a high average travel distance to the places of e.g. tree harvesting, unless cooperation over the ownership boundaries can be arranged. The cost relationships are shown generally in Figure 9.

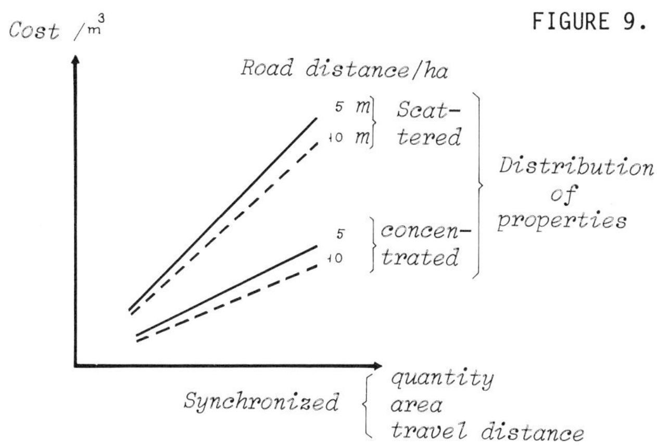

FIGURE 9. Increased concentration means increased costs of travel (commuting labour).

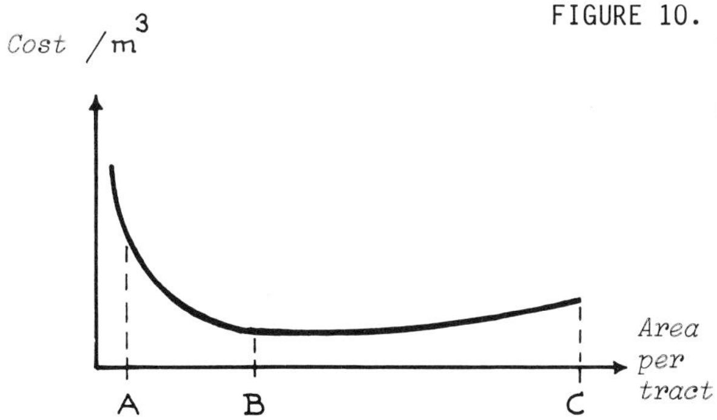

FIGURE 10. Optimum area of annual harvest on the basis of cost relationships given i Figures 8 and 9.

A summation of the curves in Figures 8 and 9 gives relatively high costs of personnel travel in Forest guard (ranger) district A. (Figure 10).

Forest guard (ranger) district B

The forest guard (ranger) district B in Figure 7 having three labour centres and corresponding labour areas represents a population situation common in the beginning of the 60's.

The evolution from situation A to situation B depends on a number of well-known factors primarily concerning the surrender of unprofitable farms, migration to urban districts, reduced labour requirements in forest road construction and mechanization due to general rationalization.

In forest guard (ranger) district B there are three villages, each one with its labour area containing five cells. The more concentrated population facilitates a relatively strong concentration in the planning of operations. The total cost of road maintenance, scaling, supervision, personnel transport and camping facilities in District B is lower than in District A. cf Figure 10.

In recent years the number of operational alternatives has increased due to intensified mechanization and the requirements with respect to organizational ability and technical skill of the personnel have increased steadily.

The law of the large numbers applies quantitatively to most forms of mechanized forms of tree harvesting. Concentration of the harvest areas and large quantities of timber reduce the direct costs.

Thus, if fragmentation of the operations can be avoided by concentrating the work, advantages are achieved with respect to technical and economic benefits. The corner-stone in planning is the labour available to the unit of operation.

Forest guard (ranger) district C.

The forest guard (ranger) district C with a very high degree of concentration in harvesting operations and with respect to population represents the situation today. A high concentration is achieved if an urban area occurs in the centre of the district and if the area of tree harvesting or the annual area is consolidated (Figure 11).

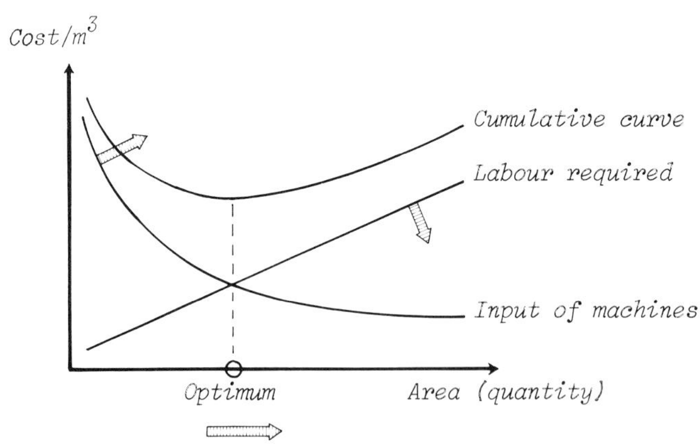

FIGURE 11. Principally, mechanization moves the lowest accumulated cost towards increased concentration.

A strong concentration is assumed to cause certain incremental costs in the form of high allowances for walking and commutation time. However, a continuous expansion of the road system will reduce travelling time from residence to place of e.g. tree harvesting and the travel costs will decline as the labour required per unit of timber will be reduced by mechanization.

The cost reducing effect of concentration on the cumulative curve in Figure 11 lowers the right part of the curve and the point of optimum is moved to the right towards a stronger concentration of the operations.

Intensified mechanization has a similar effect on the cumulative curve. The costs of tree harvesting or transport of timber per unit of volume are reduced in proportion to increased amount of timber which means that the point of optimum is moved to the right towards a stronger concentration of the operations.

A decline in the population of the forested regions has brought about an adjustment of the labour to the needs in the forests. Simultaneously there has been an increase in the population of communities and densely populated areas (urbanization). This re-grouping of the population has produced an increased concentration of the operations.

The current redistribution of forest ownerships into more compact properties has also produced an increased concentration of the operations in many locations.

Growing stock and volume of timber harvested

Another factor influencing the concentration of operations is the growing stock per hectare. In locations where the land has a low site quality increased concentration of operation is desirable. This is achieved by the harvest of large areas and by a reduced number of operations during the life time of a stand.

Quantity of thinning is a function of stand volume, form of thinning applied, and length of thinning interval.

The volume of timber removed can be increased by lengthening the interval of thinning. Thus, if the interval of thinning is increased from 5 years to 15 years, the volume of timber that can be removed may be trebled. Due to the increased volume of timber, the direct costs of harvest and terrain transport per m^3 are reduced as well as certain costs of road maintenance, tree marking and supervision.

The discussion above has shown how a relevant and important complex of problems can be identified and framed. It has been shown that i.a.

- high road net density
- concentrated population
- consolidated property distribution
- long intervals of thinning

give lower costs per m^3 for

- road maintenance
- supervision
- transport of personnel

when harvesting of timber is carried out on a few, large tracts of operation.

Attempts at improving the efficiency of expensive machines may appear futile if sufficient attention is not paid to the planning of concentrated operations.

Problems are solved in general and in detail

This outline of the forest management activities has shown how a firm grasp of the planning process is obtained. The problems can then be solved starting from the largest unit and finishing with the numerous details.

The guidelines of the master plan are essential and should be followed as closely as possible. The plan is a means, not an end. Unfortunately it is too common that the overall plan is neglected in favor of 'refinements' in some parts of the operations. Such deviations lead to losses of timber within the total unit, the forest stands not being systematically treated.

Trends

Larger units of operation. In the end of the 70's, major units of operation were treated as labour areas with a labour supply centre in the middle.

Machine investments and labour requirements are the prominent factors influencing the concentration of tree harvesting.

Increased investment in machines followed by reduced labour requirements have moved the minimum point of the cumulative curve downward and to the right (Figure 11).

Feasible concentration. An interpretation of the curves in Figure 11 leads to the following statements:

The increasingly expensive machines require a higher utilization, up to 80-85 percent. They will necessitate relatively large continuous areas of harvest and large quantities of timber. This requirement is represented by an arrow directed upward on the left half of the curve showing the projected investment in machines.

The gradually reduced labour requirements lead to declining costs of personnel with respect to travels, camps etc. This trend is represented by an arrow pointing downward on the right half of the curve showing the projected costs of labour required.

The trends interact producing a move of the position of the minimum total cost towards increased concentration as indicated by a horizontal arrow below the graph. When an optimum level of mechanization from economic points of view is achieved, the trend setting forces represented by the arrows cease to operate.

The nominal costs of tree harvesting per m^3 in Sweden remained at approximately the same level for about 20 years. Simultaneously, negotiated prices and wages increased in several countries. Still the costs of forest operations were successfully kept at an acceptable level due to intensive efforts in rationalization and mechanization. Exceptions have been the costs of labour intensive small timber and the costs of timber from remote or difficult-to-reach areas. The real costs have declined during the same period in view of the inflationary forces, i.e. by more than 50 percent.

Another factor of importance for increased concentration of the annual harvest area is the travelling time required for transports of personnel from residence to place of work. It is assumed that labour is given the benefit of living in their own homes, satisfying a reasonable social need.

The maximum travel time from the residence in a densely populated area to the remote parts of the forest operation should not exceed one hour, corresponding to a distance of approximately 60 km by personnel vehicle. Thus, the acceptable travel time could define the outer boundaries of a forest operation unit or a management area.

New forms of operation. The old forest guard (ranger) districts, which were naturally defined units of organization, were disappearing in the 70's. This change is current and it will lead to an organization of work functions with activities assigned to specialists of various kinds. Only a limited number of districts may be retained for administrative purposes.

A work crew of seven men in an old forest guard (ranger) district encompassing 10,000 hectares in 1980 meant that the need for forest guards (rangers) no longer existed.

In 1954 the optimum work crew was 66 men for the same district. Work of the forest guard (ranger) was then supervision, personnel matters and administrative in nature. Those were tasks that have been sharply reduced by the introduction of machines, the number of employees being cut considerably.

The activities within a district (forest guard or ranger) or a management unit have assumed an entirely different profile. The numerous and expensive machines, which require full utilization and careful planning, have increasingly set the pace of work in forest operations.

Planning in detail

Annual areas, units of treatment and stands will now be discussed on the basis of the situation described for district C (Figure 7).

A unit of operation is assumed to cover 10,000 hectares of productive forest land and 15 years are considered to be a feasible interval of treatment. The assumptions will mean that there will be 15 annual areas, each area approximately 670 hectares, within the forest guard (ranger) district or the unit of operation. (Staaf, 1972).

The extent and natural boundaries of the 15 annual areas can be outlined from the points of stand treatment and tree harvesting without difficulty using air-photos indoors and supplementing with reconnaissance. After the needs for stand treatments and final harvest operations have been considered in addition to an evaluation of the availability of roads and planned routes of transport, the order in which the annual areas should be treated can be ascertained. The result of this work is an operations map which may look like e.g. Figure 12 which is based on an actual case.

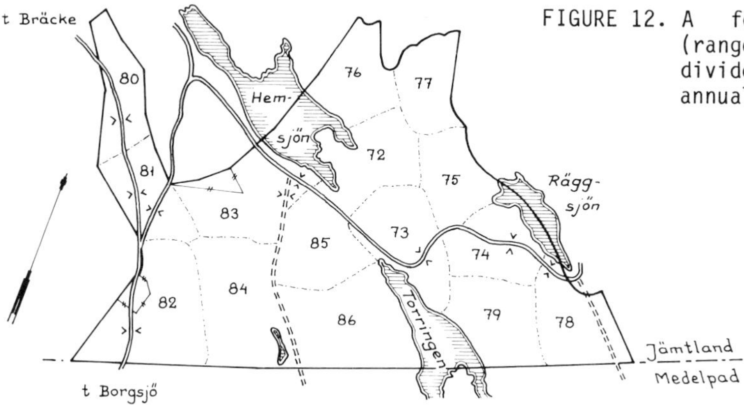

FIGURE 12. A forest guard (ranger) district divided into 15 annual areas.

73

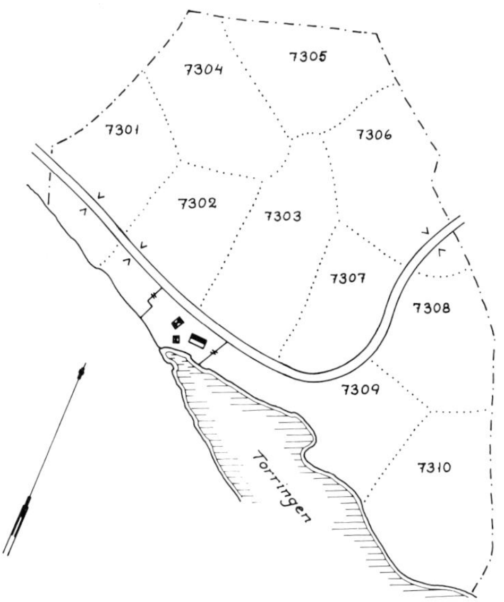

Annual harvest area No. 73

7301	Area harvested and cleared
7302	First thinning, operation in summer
7303	Second thinning, operation in summer
7304	Last thinning, operation in winter
7305	Planted stand, no harvest operation
7306	Final harvest, operation in spring
7307	Cleaning, operation in summer
7308	First thinning, operation in autumn
7309	Second thinning, operation in winter
7310	Final harvest, operation in winter

Figure 13. Annual area with units of treatment.

This work is followed by a planning of the annual area, which in this case has been given the number 73, subject to treatments in the year 1973/74.

Suitable measures to be carried out during the year are determined on the basis of air-photos and a further scrutiny of the conditions within the annual area. The various stands are identified and coupled, when feasible, into units of treatment or tree harvesting from operational points of view (Figure 13). In the present case there are ten units of treatment having an average size of 60-70 hectares.

Planning for various seasons

Time schedules for tree harvesting, forest improvement and other work are prepared for all the units of treatment (harvesting) in the annual area. Simultaneously are prepared plans concerning the use of labour, machines, means of transport, camps and other resources.

Today's mechanized work also requires a dependable machine service for efficient operation.

The preparation of a plan for timber delivery is an important step sometimes preceded by marking of trees or field inspection. The plan gives information on the quantities of various assortments, e.g. pulpwood and sawlogs, that can be delivered from a unit of treatment in a given month.

Division of the areas of treatment into parcels for felling and transport

Planning now proceeds deeper into the details concerning the areas of treatment that will be harvested. A division of the areas into parcels for felling and transport is often required. The parcels vary in size and orientation depending on the amount of timber to be removed, form of felling, thinning or final harvest, method of harvesting, tree length timber or assortments (short timber).

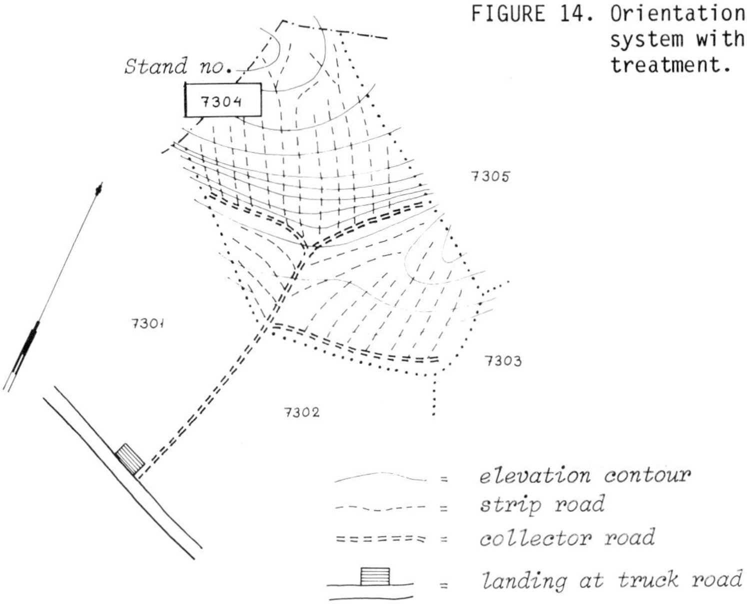

FIGURE 14. Orientation of the road system within a unit of treatment.

Road systems and other routes of transport

A large part of the harvesting work consists of transports such as forwarding and further transports. Planning of harvesting operations includes the very important planning of road systems and other routes of transport. For this purpose may first be prepared an overview of the forest road plans which contain proposals on a framework of road systems adapted to the needs of transport in the forests within a large area.

The harvesting plan normally contains a plan showing the optimum road net system. Since the costs of timber transports constitute a large part of the total cost of production, it is essential in the forest operations to reduce the costs of transport by means of a rational system of roads and other transport routes. In addition to roads for heavy truck transports there may also be rivers and railways available as routes of transport over long distances.

Planning of roads

An important step in the planning process concerns the road system within the unit of treatment. Figure 14 shows how such a system can be designed in detail within the treatment unit 7304.

A number of more or less parallel strip roads are the outermost branches of the road system connecting to the collector roads. The collector roads converge towards the access road of the annual area e.g. a road for heavy truck transport. At the points where the collector roads connect to the access road, planning has provided for a timber landing of a size suited to the need for a buffer pile between forwarding by means of tractor and further transport to industry by means of trucks.

Environmental features influencing tree harvesting

General features of environment

The forest environment is a result of the interaction between a large number of factors of a geomorphological, climatological, geological and ecological nature behind which the sun is the original source of energy.The environmental features can be classified and put into systems in several different ways. To varying degrees the features influence the primary and secondary production in forest operations. In the primary production the environmental features regulate i.e. the composition of the plant society with respect to trees, bushes and ground vegetation and its vigor expressed in density, amount and height.

In the secondary production the environmental features influence in various ways people, animals and equipment.

Increased mechanization has given people e.g. the machine operator enclosed in a cabin, an artificial work environment. The performance of the machines is influenced by the environmental features which in turn are affected by the machines.

A brief presentation will be given concerning the most important environmental features in forest operations: geographic location of the work area, climate features, terrain and the forest stand.

Geographic location and extent of the work area

Geographic location is usually defined by latitude and longitude expressed in degrees. Length of daylight and seasonal variations are influenced by the geographic location.

Altitude of the work area in relation to the environment is expressed in metres above sea level. Altitude affects i.a. density of air and, hence, the oxygen intake of people, efficiency of the combustion engines and the lifting power of a helicopter.

Extent of the work area is often given in proportion to the total area of land and water and expressed in e.g. hectares or square kilometers, acres or square miles. The area can be continuous or discontinuous as in an archipelago. The forested area in proportion to the total area is often given in percent. The land area also includes cultivated land and waste land such as bogs and mountaineous barrens.

The distribution of the forest land by ownership varies with respect to both area and configuration.

All these features are of importance for the planning of operations and the management of our forests.

Climate features

Air temperature

Temperature conditions in the air space close to earth up to an altitude of 10 km at intermediate latitude are of decisive importance for weather. Air temperature, therefore, is one of the most important meteorological factors. (Anon. Focus Materia, 1965).

Of statistics on temperature obtained on the basis of observations compiled by official institutes, the statistics on the diurnal maximum and minimum temperature are valuable information for timber harvesting activities. The limits within which temperature varies are more important in this context than e.g. diurnal mean temperature of temperature at a given hour.

Precipitation

The annual precipitation varies largely between different regions of the earth. Amount and distribution of precipitation throughout the year is, like temperature, of great importance not only for the primary production but also for the secondary production i.e. tree harvesting operations.

Information on the variations of precipitation is very useful when choosing the size of mobile machines for year-round operations. The amount of precipitation is of importance for the choice of tree harvesting methods and equipment in general. Data on the frequencies of temperature and precipitation etc, therefore, must be considered when planning harvesting operations, long term or short term.

Winds influence the felling of trees.

Wind is influencing the felling of trees in particular. At brisk wind, having a force of 6 Beaufort, corresponding to a wind velocity of 10.8-13.8 m per second, tree felling work is made very difficult, particularly when felling is to be directed. High winds with a force of 7 Beaufort, corresponding to a wind velocity of 13.8-17.1 m per second, probably prevents felling altogether.

Wind felling

Gales and hurricanes with a wind velocity exceeding 24.5 m per second are usually inflicting severe wind felling which can devastate large forest tracts.

In November 1969, for instance, high winds caused heavy damages in the form of wind felling in middle Sweden. During the storm with gusts up to 24-30 m per second, approximately 5 percent of the growing stock or 400,000 m^3 of timber equivalent to three annual harvests were felled in a forest district (Figure 15).

Wind felling of that magnitude creates difficult tree harvesting problems and spoils the plans. Windfelled timber, which must be salvaged quickly in order to prevent decay and insect damage, requires drastic measures from a technical point of view.

The forests within a windfelled area are often totally damaged either by entirely levelled stands or windfelled groups of trees and scattered trees, which largely affects the primary production. Large reforestation projects must be undertaken, stand boundaries re-established and rehabilitating stand improvement measures carried out throughout the whole area that has been damaged.

FIGURE 15. Tree harvesting after windfelling.

Harvesting of timber felled by wind has been encumbered by many accidents several of which have been fatal.

Wind velocities exceeding 10 m per second complicate or prevent transports by means of helicopter or small aircraft.

Terrain features

Classification of terrain

Terrain conditions have a great influence on the choice of harvesting techniques and on the costs of tree harvesting. A classification of the terrain conditions, therefore, is of great value when planning the harvest operations and choosing the techniques of tree harvesting. Depending on the extent of the harvesting operations, terrain can be classified on a large scale, macro-classification, and/or in detail, micro-classification.

Macro-classification is applied when whole mountain (hill) sides or large uniform plains or regions are the object of harvesting. Macro-classification will then be considered as an average description of the area. This type of classification is useful for a general planning of forest operations on a large scale e.g. whole management units.

Dominant terrain features

Whether the classification of terrain is of macro- or micro-scale, there are primarily three terrain features which are of great importance for the tree harvesting work: Carrying capacity of the ground (ground conditions), configuration of the ground surface (surface structure), and slopes. Classification schedules for practical purposes have been developed for a rating of the terrain conditions on the basis of the dominant terrain features.

Terrain is here considered to be a tract of land without population, houses or other permanent establishments such as e.g. traffic routes of various kinds, or in other words, a natural landscape with forests, swamps, mountains etc.

Carrying capacity of ground

The carrying capacity of ground, i.e. its capacity to resist physical pressure, primarily depends on type and moisture of the soils. For practical purposes there is no carrying capacity when e.g. a vehicle cannot proceed because of bogging down. The carrying capacity is complete when ground leaves no traces of pressure. The carrying capacity is usually expressed in N (Newton) per cm^2.

A high moisture content makes fine grained soils loose, hence, creating a low carrying capacity. A contributing cause is often the occurrence of heavy humus formations. Such water retaining soils are called cohesion soils.

Completely dry soils usually have a high carrying capacity. Soils consisting of coarse materials such as gravel and sand, which provide good drainage, belong to this category. These soils are often called friction soils. The carrying capacity varies between these extremes when ground is not frozen. It is also affected by the occurrence of soil binding (reinforcing) material such as roots, ground vegetation, rocks and boulders.

According to the soil pressure theories applied within the house construction technology, the distribution of load on friction soils is subject to calculations. (Scholander, 1973).

Surface structure

Surface structure in terrain has a great influence on cross-country (off-road) transports and other work involving people, animals and machines in tree harvesting.

The great variations in the surface structure in terrain depend on factors of unorganic or organic nature. The obstacles they constitute with respect to work activities depend on the nature, size and number of the obstructions. From a practical point of view the obstacles can be grouped into minor obstacles and major obstacles.

Obstacles on the ground have a more or less negative influence on the efficiency of work involving trees and on cross-country transports to or from a given sector of terrain.

Minor obstacles

Minor obstacles such as stumps, windfelled trees, bushes, boulders, trenches, creeks, hollows and mounds may often be passed straight over by machines which have sufficient clearing space. It is easy to recognize the cost of e.g. time lost when a tractor load is stuck on a stump.

Major obstacles

Major obstacles must be by-passed. A high frequency of boulders in an area can prevent entirely the operation of wheel tractors or even caterpillar tractors. The technical solution for a harvesting operation in such an area may be the application of some winch system.

Statistics on micro-terrain features

In recent years the national forest inventories in Sweden and Norway have collected statistics in combination with data on timber for the purpose of enabling regional and national analyses and planning of tree harvesting activities. Being of great importance for the performance of e.g. cross-country (off-road) transports, slopes have been given special attention.

Schedules for classification of terrain types may contain primarily three basic features which are judged independent of each other, e.g. slopes, ground conditions and surface structure. The nature of schedules for terrain types will vary from one country and location to another depending on the particular geological formations, sites, obstacles etc. However, some simple examples from Sweden may be given. (Skogsarbeten, 1969).

Slopes

Sloping ground surfaces with an area exceeding 15 m^2 (space occupied by a harvesting machine) is an occurrence (macro-obstacle) which can be considered as a separate terrain factor subject to classification.

Slope of the ground surface in relation to the horizontal plane varies greatly from one area to the other, or from one country to the other. Compare Norway, the Alpine countries and Western North-America, which have a very high occurrence of sloping ground, with Sweden, Finland and the relatively flat country of Denmark.

Slopes, up-hill, down-hill or side-hill, have a great influence on cross-country movements and (off-road) transports in terrain (Figure 16). The most common way of expressing a slope is to give the difference in elevation between two points in percent of the horizontal distance.

Slopes can also be given in degrees or by difference in elevation expressed in percent of the slope distance between the points.

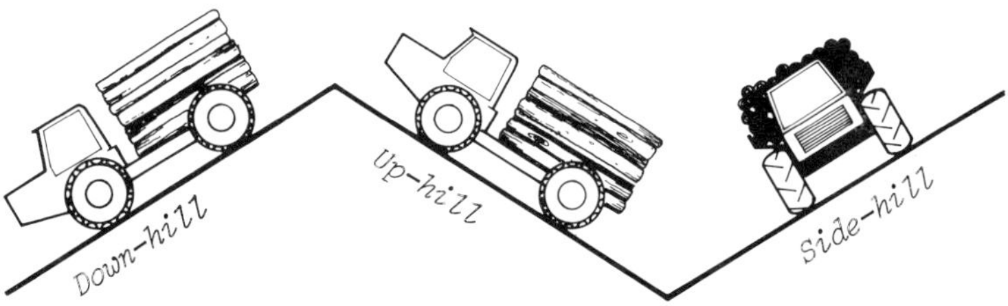

FIGURE 16. Various types of slopes.

In a terrain type schedule the slopes may be divided into five classes or intervals expressed both in percent and in degrees (Table 1).

Table 1. Classes of slopes

Class	Percent	Degrees
1	0 - 10	0 - 6
2	10 - 20	6 - 11
3	20 - 33	11 - 18
4	33 - 50	18 - 27
5	50 -	27 -

Ground conditions

In this context ground conditions is an expression of carrying capacity which may be divided into five classes such as described below.

Class 1 Very good ground conditions
Class 2 Intermediate class
Class 3 Average ground conditions
Class 4 Intermediate class
Class 5 Very poor ground conditions

Surface structure

Surface structure, which is characterized by the occurrence, height and nature of obstacles (e.g. boulders and stumps), may also be described by five classes.

The trees

Since the choice and performance of tree harvesting equipment depends strongly on the size and shape of trees, length of useful timber, length and structure of the tree crowns, the following tree characteristics may be mentioned briefly:

- Diameter
- Crown
- Limbs
- Weight of wood and bark

The tree statistics listed above can be obtained from sample tree material collected separately for each location.

Diameter

The diameter of tree trunks varies with species, age of tree and quality of the site. Figure 17 shows a common relationship between the cost of harvesting per m^3 of timber bunched at the stump, and the diameter of the tree trunk. (Sprängare and Troedsson, 1970).

Thus, it is desirable from the point of tree harvesting to produce big trees preferably of uniform size in the future stands that will be thinned or finally harvested.

As long as the trees are processed individually, it is desirable from a harvesting point of view that the variation of the sizes of felled trees is reduced. Size of the machines can then be easily chosen and utilized properly.

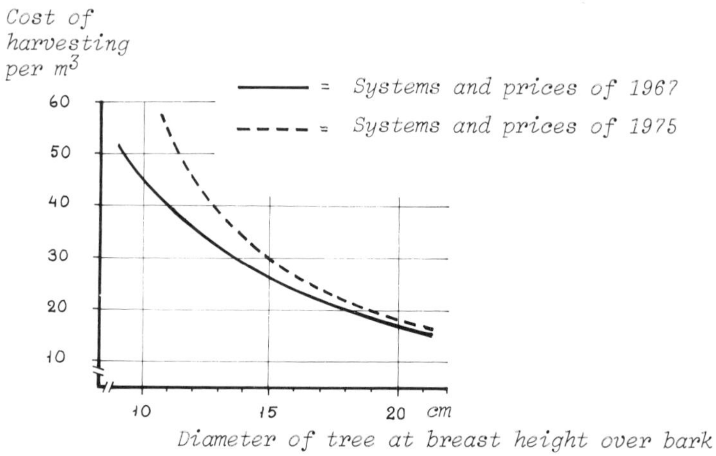

FIGURE 17. Influence of tree size (Dbh) on the cost of harvesting (volume of timber removed 50 m^3/ha) by a conventional assortment method.

Crowns

Information on the length of the tree crowns is of great importance for a judgement of the processing difficulties. The distributions of total tree height and height to crown base in various diameter classes are here of special interest.

Crown width and diameter of the biggest trees show a relationship of special interest. The difference between species in this respect is of importance for delimbing operations.

Limbs

In the field work of a national forest inventory the coarsest limbs were measured according to the following instruction: "Measurements of the coarsest limbs are to be taken from all sample trees bigger than 20 cm over bark at breast height. Measurements are to be collected from the useful section of the trees. Diameter of the limb is to be measured approximately 5 m from trunk (outside the swell) and recorded in cm to the nearest lower unit. Measurement is taken on the smallest diameter.

Measurements are taken in order to obtain an approximate idea of the frequency of trees with extremely coarse limbs. Knowledge of the largest diameter of the limbs is necessary for the choice of cutting tools in a delimbing machine."

Weight of trees

Mechanization has increasingly made whole trees or tree length trunks the units of handling in harvesting operations. Knowledge of weight and centre of gravity in these relatively large objects of work is necessary when methods and technical means are to be chosen.

The basis for an estimate of average and maximum weight of whole trees and trunks can be obtained after a number of investigations and measurements.

Some examples of calculated average weight of whole trees (trunk + crown) and trunks of spruce are shown in Figure 18.

Since weight of whole trees varies greatly within the same diameter class depending on shape of trunk, tree height, density of wood etc., it is of value to know the maximum weight (Table 2).

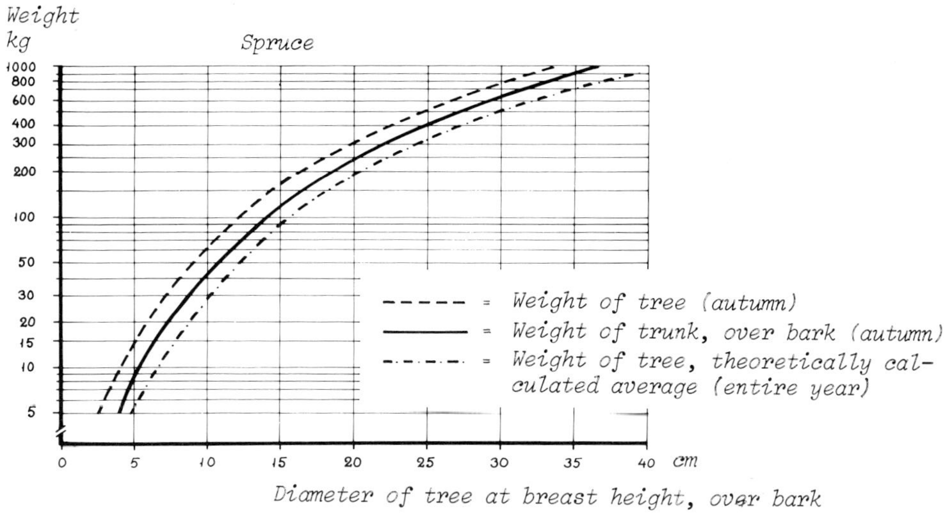

FIGURE 18. Weight of trees and trunks of spruce (Ager et al, 1964).

Table 2. Maximum weight, percentage higher than average weight, of whole trees (trunk and crown). (Ager et al, 1964).

	10	20	30	40	50	Diam. over bark
Pine	50	45	40	35	30	
Spruce	60	55	50	45	40	

Density of wood, center of gravity in trunks and statistics on bark

Table 3 gives some examples of the average density of trunks including or excluding bark in late autumn.

Table 3. Average density of recently felled timber of pine, spruce and birch expressed in kg per m^3 solid wood.

Species	Region	Average density, kg/m^3 Incl. bark	Average density, kg/m^3 excl. bark
Pine	Southern Sweden	910	960
	Northern Sweden	860	890
Spruce	Southern Sweden	890	900
	Northern Sweden	800	810
Birch		900[1)]	900[1)]

1) at time of sap flow 1000 kg.

The average form quotient of a tree may be approximately 0.65. Calculations have then shown that from a practical point of view the centre of gravity is located at 40 percent of the tree height from the stump. Knowledge of this relationship is of interest at calculations of traction required for skidding of trunks or trees.

Knowledge of the volume and weight of bark is of interest primarily in transport calculations. Data on the volume of bark can be found in forestry handbooks showing e.g. that volume of bark varies between 40 percent and 9 percent in trees with a diameter ranging from 6 cm to 56 cm at breast height, respectively.

Small trees have a relatively large volume of bark. The same applies to weight of bark as shown in Figure 19 for spruce. (Jonson, 1929).

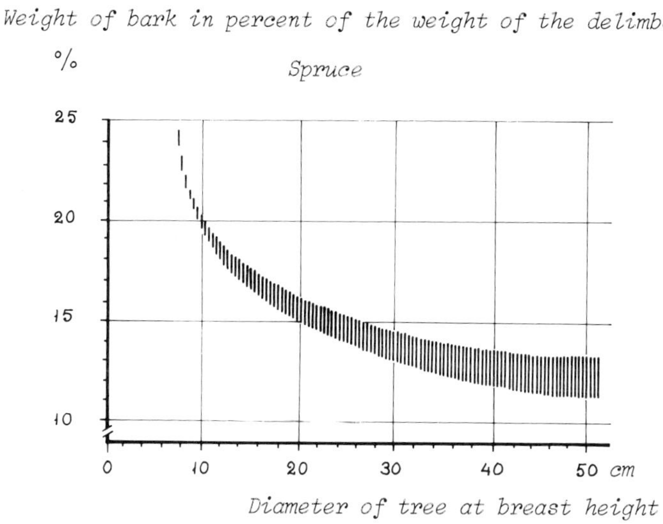

FIGURE 19. Weight of bark in spruce in percent of the weight of the delimbed trunk. (Jonson, 1929)

Units of timber handling

Trees vary widely with respect to weight and quantities per unit of area. Figure 20 shows how weight varies in pine between approximately 15 kg for a 5-cm (Dbh) tree and 2000 kg for a 50-cm tree. Simultaneously, the volume of timber harvested may vary between 10 m^3 and 500 m^3 per hectare. This variation has resulted in units of timber handling ranging between chips and whole trees. (Staaf, 1965).

The unit of timber handling shows a principle relationship in Figure 20. The relationship largely depends on the availability of technical resources, primarily for lifting and hauling adjacent to the stump.

FIGURE 20. Principal relationship between size of tree and unit of timber handling

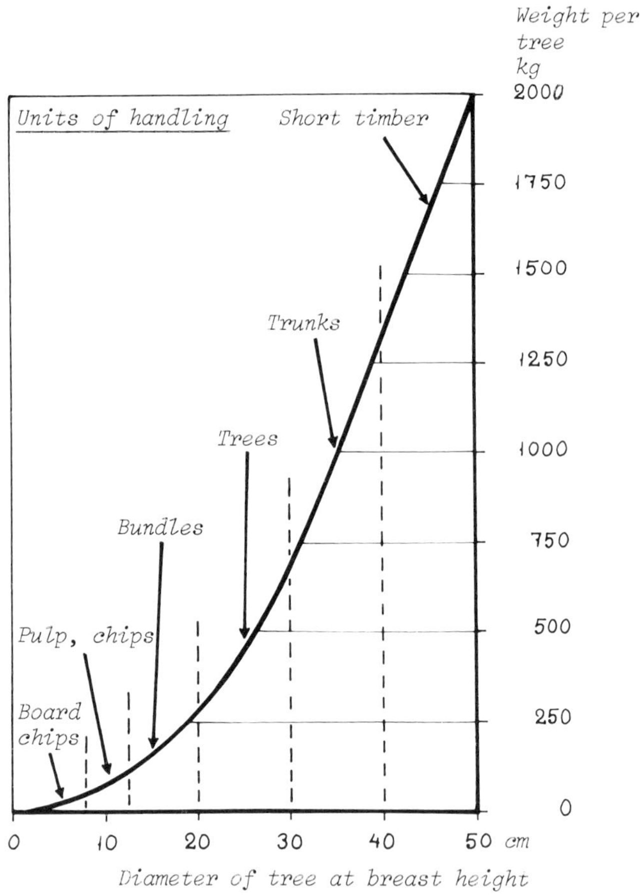

The forest stands

At the choice of equipment and the design of methods for tree harvesting operations, knowledge of the composition of the stands with respect to volume and weight is an important prerequisite.

In mechanized forest operations the capital cost is a major item. The cost of a machine is strongly dependent on the size of the machine which in turn is adapted to the work involved in timber processing, handling and transport.

There is a general relationship between the tree sizes, project sizes and sizes of machines. Knowledge of volume per hectare, weight of trees, size of limbs and other statistics are also important for the proper construction and design of machines used in forest operations. Data on stands and trees can also be used in other activities in the forests e.g. calculations of labour requirements and planning.

Relationship between cost of harvesting and volume of timber

Since the cost of harvesting or labour requirements depend on the volume of timber removed per hectare, this relationship should be taken into consideration at the planning stage. The relationship between cost of harvesting and the volume of timber removed per hectare is shown in Figure 21. (Sprängare and Troedsson, 1970).

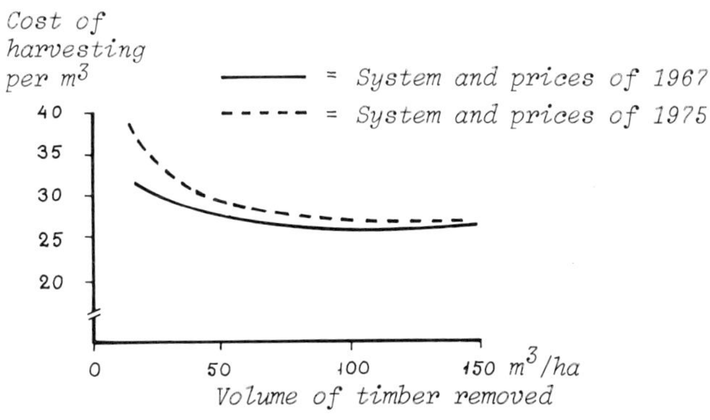

FIGURE 21. Relationship between cost of harvesting and volume of timber removed per hectare according to the systems and prices of 1967 and 1975 for an average tree diameter of 15 cm.

Mobility in the stand is an important environmental feature. This applies in particular to thinning. The denser the stand the more difficult is felling and processing of trees and bunching of timber.

Thinning

In a stand mature for the first thinning, crowding of the trees is pronounced and profit of the harvest operation is low. This is a major problem in our forests today. Harvesting of small trees requires a high labour input in relation to the yield of timber. However, thinning is necessary in order to produce a desirable diameter growth of the trees within a reasonable time.

The highest total timber production per hectare is achieved in unthinned stands. However, such stands are affected by losses from natural mortality, and harvesting of the relatively small trees is very expensive.

Clearcutting

At clearcutting, which is a common form of final harvest, there are usually no problems of crowding in the stands except in certain cases with directed felling applied in order to serve delimbing machines.

Tree harvesting techniques

We have now discussed briefly terminology, forms of production in forestry and some of the background conditions which influence planning of the harvesting operations viz. where and when the trees and the stands should be harvested in an orderly fashion.

We will now proceed into descriptions of the various partial operations that can be distinguished in tree harvesting. The order in which they are discussed does not necessarily represent the sequence in which they will be applied after the method of tree harvesting techniques has been chosen on the basis of the particular circumstances, equipment and labour available.

Partial operations

Harvesting and associated transport of timber from the stump to the mill comprises a number of partial operations, time and place of which in the harvesting process can be varied in several ways. For natural reasons felling is the only operation which must be done in a definite place, the tree site. Other operations can be carried out in various places between the tree site and the mill.

Nature and placement of the various partial operations in the harvesting sequence is determined by the means of harvesting, transport, and technology that are judged to be the most feasible ones for the particular area of operation. (Skogsordlista, 1969).

The following partial operations will be treated in the subsequent chapters: felling, delimbing, bucking, debarking, bunching and chipping.

Felling

Felling is the partial operation which dominates tree harvesting work. It includes measures undertaken in order to separate standing trees from the stumps or roots, or other methods necessary to fell the tree. Although felling requires a relatively small proportion of the total time involved in the harvesting of a tree, its proper performance is of importance for the subsequent processing.

The term processing is often used for the preparation of the felled trees for transport i.e. delimbing, bucking-scaling, debarking and chipping.

In some harvesting systems the trees can be processed without first being felled to the ground.

Methods of felling trees may be described according to various systems. The following systems are based on the methods of treating the roots and the direction in which the trees are removed from the sites. (Staaf, 1972).

1. Harvest of trees without roots

1.1 Trees are separated from the roots (stumps) and felled into horizontal position

1.1.1 Trees are felled before being processed in horizontal position

1.1.2 Trees are felled after being processed in vertical position

1.2 Trees are separated from the roots (stumps) and lifted in vertical position

1.2.1 Trees are lifted for processing in horizontal position

1.2.2 Trees are lifted for processing in vertical position

2. Harvest of trees with roots

2.1 Trees are harvested with entire root systems

2.1.1 Trees are felled with entire root systems

2.1.2 Trees are lifted with entire root systems

2.2 Trees are harvested with cut root systems

2.2.1 Trees are felled with cut root systems

2.2.2 Trees are lifted with cut root systems

Choice of felling object

The object of felling, the tree, is selected in the stand, often being marked at least in thinning operations when only certain trees are to be removed. At final harvest operations, usually all trees, excluding in some cases seed trees, are felled.

Felling year-round

Felling is now carried out largely as a year-round operation. Thus, felling and processing is done in any season under almost any climate and weather conditions.

Direction of felling

Felling is actually a first transport step in a direction which the feller or machine operator can control by a skilful handling of equipment. If felling is done in the right direction, a valuable free transport can be obtained by gravity.

Since a tree can usually be felled without difficulty in almost any direction within the 360° circuit, the direction of felling should be established before harvesting starts.

Felling of trees can be done in a random direction, which is applied at so-called rush felling, where trees are felled criss-cross without any thought given to the subsequent transport. This form of felling was common when timber was hauled by horses.

Directed felling

Modern felling is usually done in a predetermined direction, so-called directed felling. This form of felling also facilitates a concentration of the timber to predetermined places in the stand or along strip roads for cross-country (off-road) transport.

At directed felling the butt ends and the top ends of the logs (or trees) are placed in a pattern which facilitates the subsequent operations. Conventional felling and processing by means of chain saws can be carried out by directed felling of trees over some previously felled trees in order to arrange for delimbing and debarking at a feasible working height (50-70 cm above the ground). Simultaneously, the shortest possible distance of subsequent bunching is obtained.

Terrain conditions, which determine the direction of transport, are also of primary importance for the direction of felling. It is, therefore, convenient with respect to felling and the establishment of an optimum direction of felling if a strip road system for the short terrain transport of timber has been laid out and marked within the area of harvest prior to felling.

Planning of work within a harvest area should take into consideration the common changes of wind direction in order to provide for a shifting of felling sites at strong winds. Felling against the wind is very heavy work, it is less precise, and hazardous.

Tools and means of felling

Old tools. The old tools for felling of trees included axes, log saws, 1-man saws, 2-men saws and bow saws. These tools are now of historic interest only. However, they may be considered as steps of development towards modern tools.

Chain saws. After the second World War the portable chain saws were introduced as tools of felling. Initially heavy 2-man saws (weighing up to 40 kg) were introduced but they were subsequntly replaced with light one-man saws, today weighing approximately 4-7 kg at an engine output of approximately 3 kW.

The chain saw is now an indispensible tool in the forests, not only for felling of trees but also for delimbing and bucking. Proper techniques at work with chain saws reduce the risk of accidents caused by fatigue, requiring less physical strength and giving better precision and improved work output.

Working and holding positions

Advice concerning working positions should be followed e.g. with respect to position of the feet. A straddling stand gives the operator a better balance than that given by closely placed feet. Since a lowering of the centre of gravity also improves the sense of balance, the operator should keep the body in a low position.

Proper holding is important. The thumb should be held underneath the handle bar in order to prevent 'unnecessary' accidents.

The disadvantages in using chain saws with respect to noise, vibrations, emissions and hazards, e.g. throws, are treated in books on ergonomics, machine technology and worker protection (ILO, 1981).

Preparations

Certain preparations are made prior to felling, e.g. walking towards the tree with engine idle and directed forward, removal of undergrowth, which may interfere with felling, and delimbing of the tree trunk up to breast height.

Guiding cut and felling cut

Proper work procedures when making the guiding cut and the felling cut are very important.

The guiding cut, consisting of an upper and a lower cut, guides the direction of felling which, as mentioned above, is of importance for the subsequent work phases. The guiding cut should have a depth of approximately one-fourth of the diameter of the trunk.

The upper and lower cuts are made in big trunks with saw chain pulling, in small trunks with saw chain pushing. The guiding cut should have a 45° opening (Figure 22).

When the guiding cut is made, direction of felling is controlled, sometimes aided by means of a lever.

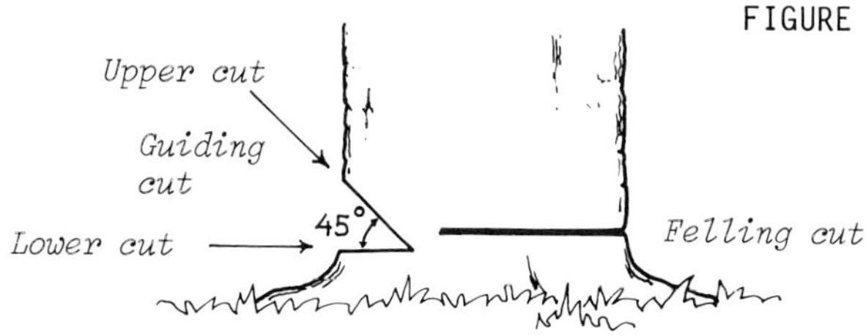

FIGURE 22. Guiding cut with upper and lower cuts is usually made by means of a chain saw.

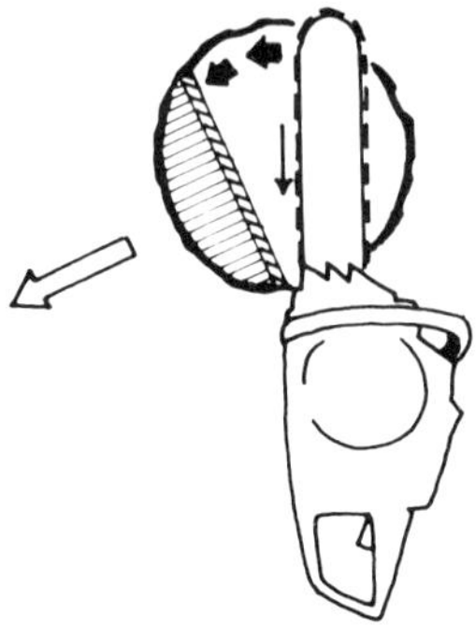

FIGURE 23. Application of chain saw to small trunks.

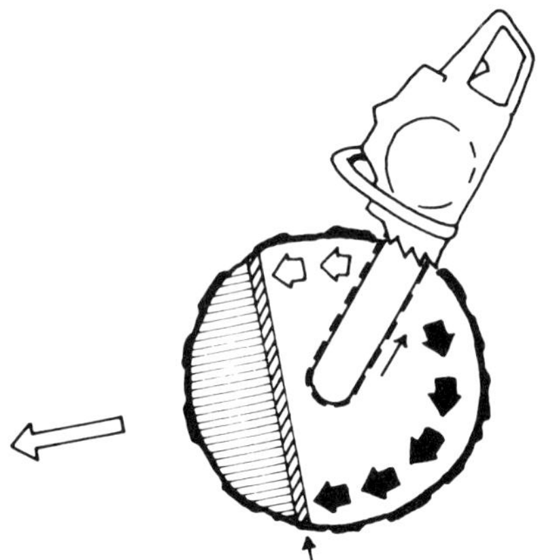

FIGURE 24. Application of chain saw to large trunks.

The felling cut is made slightly higher than the opposite guiding cut (maximum 3 cm) in order to facilitate felling.

At felling of small trees the saw chain should be pulling after being applied at a point approximately 10 cm from the inner part of the guiding cut. This point is then used as a hinge for sawing towards the guiding cut until approximately 3 cm remain for the breaking crest.

At felling of large trees the saw blade is first inserted approximately 10 cm from the guiding cut. The saw blade is then pushed towards the guiding cut until a feasible breaking crest (approximately 3 cm thick) is obtained. The saw is now brought around the tree until an equal breaking crest has been obtained on the opposite side of the tree. See Figures 23 and 24.

After the guiding cut and the felling cut are finished as above, the tree is brought to fall by pushing or by means of a lever.

Some safety rules at felling

- use proper equipment
- keep distances, double tree length to nearest co-worker

- never go underneath jammed or stuck trees
- clear undergrowth before felling work starts
- make a correct guiding cut
- leave a breaking crest
- keep thumb under the front handle bar
- keep close contact with the chain saw
- work with bent knees and feet apart
- stop the engine or use chain brake when moving to the next tree
- Make the felling cut slightly above the inner part of the guiding cut. The heel formed in the end surface prevents the tree from sliding backward over the stump.

To fell a tree in the desired direction use a felling wedge, breaking lever or a felling pad into which air can be pumped by means of the chain saw.

The felling lever has a built-in leverage ratio of 30:1 i.e. a lift of 100 kg on the handle gives a lifting power of 3000 kg on the tree. Lift correctly, straight back and bent knees.

If lifting height on the handle is insufficient to bring the tree to fall, make a swift "retake" by pushing in the lever when the tree is moved as high as possible until the butt end is resting against the upper plate.

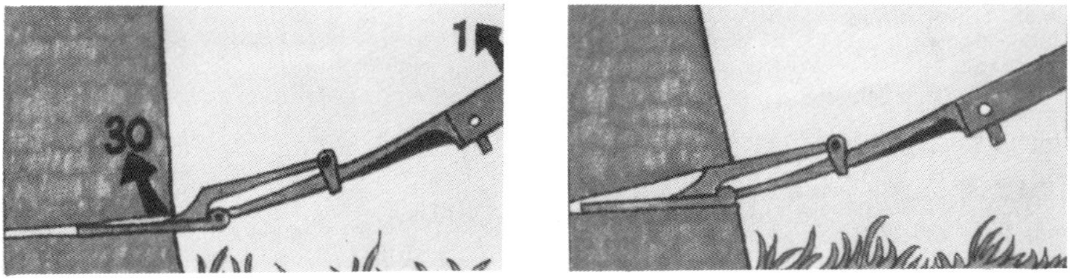

FIGURE 25. Double felling lever.

Use of felling pad

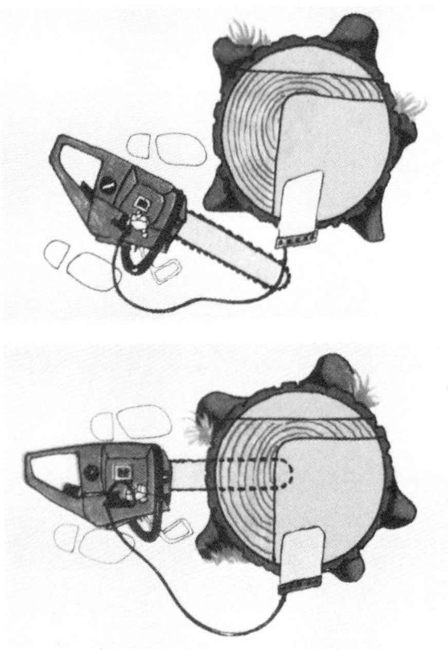

Move from the chain saw to the other side of the tree and insert the felling pad as shown in the picture. The pad should be turned so that the attachement of the hose is close to the chain saw. Connect the hose to the saw, prime the engine and push the vent button for a couple of seconds.

Saw the second part of the felling cut obliquely underneath and slightly overlapping the first part. Push the vent button during the last 5-10 seconds of the felling cut. The pad will then expand and the tree will fall.

FIGURE 26. Felling pad.

Clipping and shearing tools. In the 1960's hydraulically clipping and shearing mechanisms were developed for felling and bucking.

Clipping tools. The clipping mechanisms are designed with two components that work according to the principle of scissors or a double guillotine. See Figures 27 and 28.

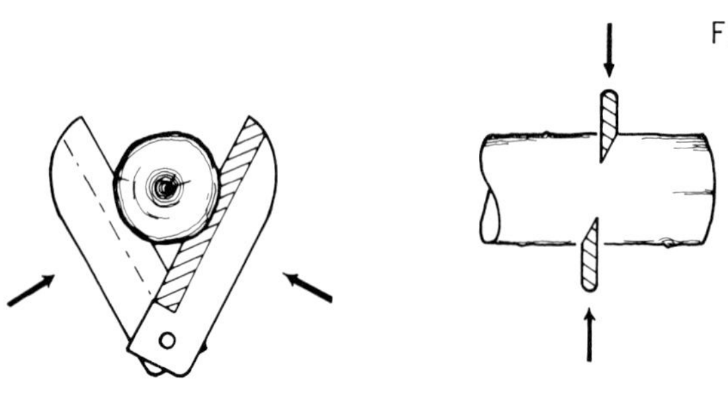

FIGURE 27. Hydraulic clipping tool designed as a double acting pair of scissors.

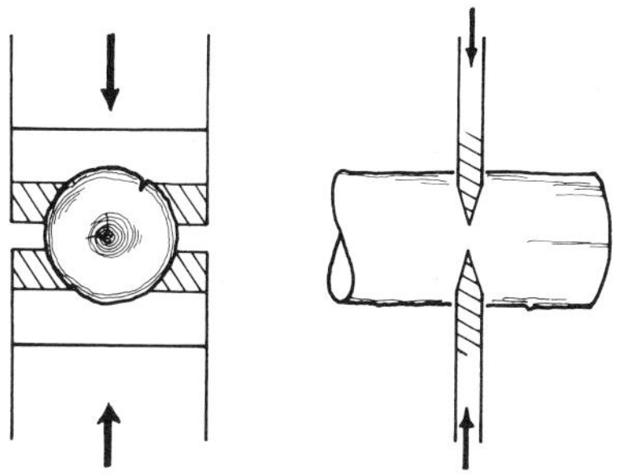

FIGURE 28. Hydraulic clipping tool designed as a double-acting guillotine.

Harvesting machines equipped with clipping tools for felling are i.a. a felling machine with double acting guillotine and a processor designed for bucking.

Shearing tools. The shearing mechanisms have a working component with a counter support on the opposite side of the trunk. There are two types of shearing tools: one type with jointed components and one type with single acting guillotine. See Figures 29 and 30.

Several types of shearing tools have been developed for felling and they have been mounted as auxiliary equipment on caterpillar tractors and wheel tractors.

A shearing tool (single acting guillotine) mounted on a crane was first designed by the Institute of Forest Techniques at the Faculty of Forestry (Sweden) in 1966. It was given the name Garpnäven (The Garpen Fist) and it has been developed for practical purposes on a feller (Dahlin, 1966).

FIGURE 29. (right) Hydraulic shearing tool designed as a single-acting pair of scissors.

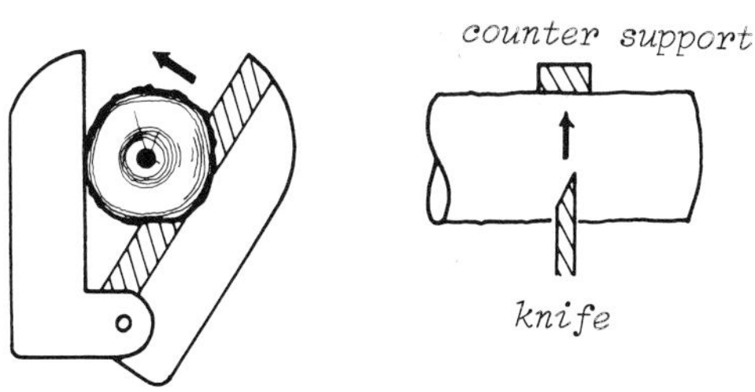

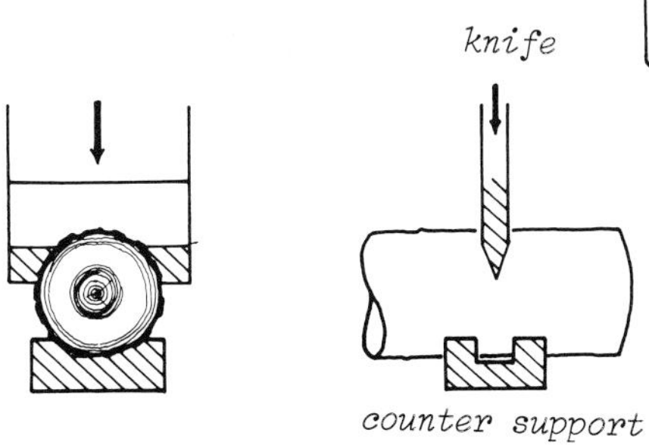

FIGURE 30. (left) Hydraulic shearing tool designed as a single-acting guillotine.

Calculation of shearing forces

Experiments have given support for calculations of the shearing forces required (Kempe, 1967).

Shearing force can be calculated from the following formula:

F = D(155s + 50D + 700) (0.2 + 2 p) - t(34D + 1000), where

F = shearing force expressed in N (Newton)

D = diameter of cross section under bark expressed in cm, perpendicularly to the directon of cut

s = thickness of tool expressed in mm

p = dry density of timber expressed in g per cm^3

t = temperature of timber expressed in °C when $t \leqq 0°C$ (note sign), if $t > 0$ insert t = 0.

Design of the formula allows an interpretation of the influence of various factors. The first parenthesis contains three terms of which 155 s is a so-called displacement force, 50D is a friction force between the tool and the working pressure and 700 is force at edge or cutting force.

These three terms are calculated per cm of edge and are then multiplied by diameter D which is equal to the maximum length of the edge.

If the relative magnitudes of the three terms are studied, it is possible to recognize the reduction in shearing force that will occur when e.g. the coefficient of friction is lowered. The whole factor contained in the first parenthesis of the formula is then multiplied by a correction factor for dry density. The value of this correction factor deviates from 1.0 if dry density of the timber deviates from 0.4 g per cm^3.

The last term in the formula gives an additional force required if temperature is below 0°C in the felling cut (Figure 31).

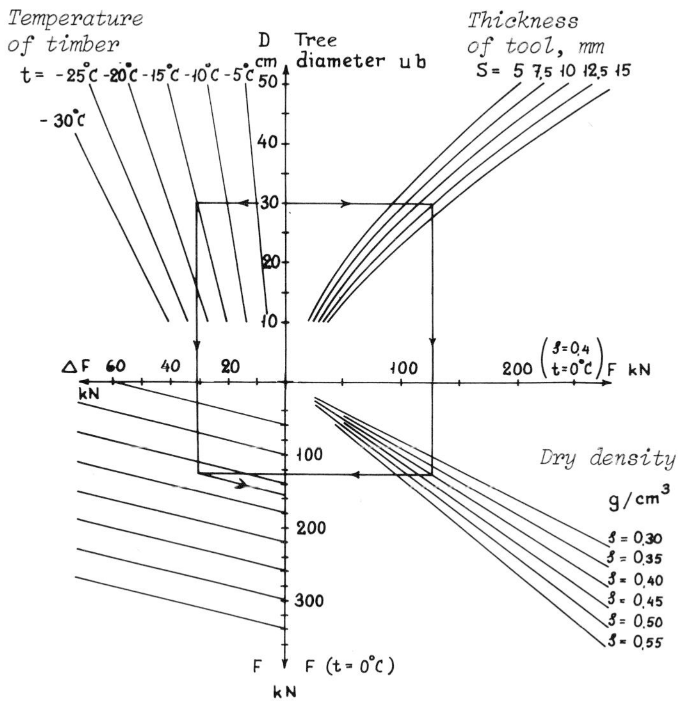

FIGURE 31. Nomogram for determination of shearing forces required at felling by means of clipping or shearing tools (Wiklund, 1967).

Cracking caused at felling by means of clipping and shearing tools

Cracking caused by the tools in the end section of timber have been an obstacle to a rapid spread of shearing mechanisms designed for felling of sawtimber. Similar to the forces required for clipping or shearing, the extent of these defects is strongly dependent on temperature of timber.

The nature of the wood defects also depends on whether severance has been done by means of double-acting or single acting felling mechanisms.

While cracking affects the value of sawtimber, it is of no major consequence for the value of pulpwood.

If the tools are given a curved form, it is expected that pressure will be directed downward into the stump and the cracks will then be reduced.

FIGURE 32. Hydraulic felling saw.

Felling saws

Instead of hydraulic clipping tools, hydraulic felling saws (chain saws) have been used for some time in American harvesters. A felling saw has been developed which does not cause any cracks in the timber (Figure 32).

Acceptance of the felling saws depends on whether their high costs of operation caused by wear of chains and blade can be balanced by the yield of timber due to less defects. Stump height is acceptable with both felling saws and felling shears.

Circular saws for felling

Felling by means of circular saws has long been of interest only in the context of mechanized cleaning. In contrast, felling of relatively large trees by means of circular saws has appeared too cumbersome because of space required. An acceptable stump height has also been difficult to maintain. However, in recent years new felling units (heads) with single or twin circular blades have been put to use in Canada and U.S.A.

Feller-buncher with circular saw

A new type of feller-buncher was introduced on the market in the beginning of 1982. It is manufactured by a firm in Quebec, Canada.

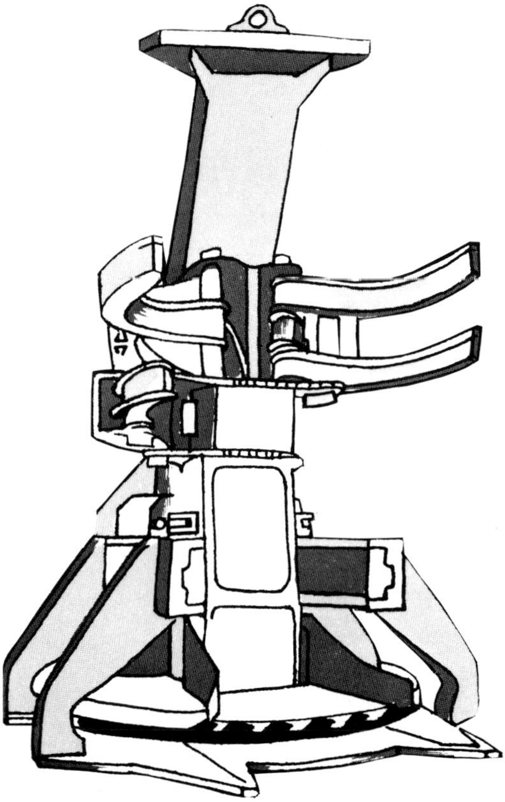

FIGURE 33. Circular saw feller-buncher.

This new felling head is a uniquely built-in circular saw that can be used to fell trees very quickly without damage to the timber or to the saw. Figure 34 shows the working principle.

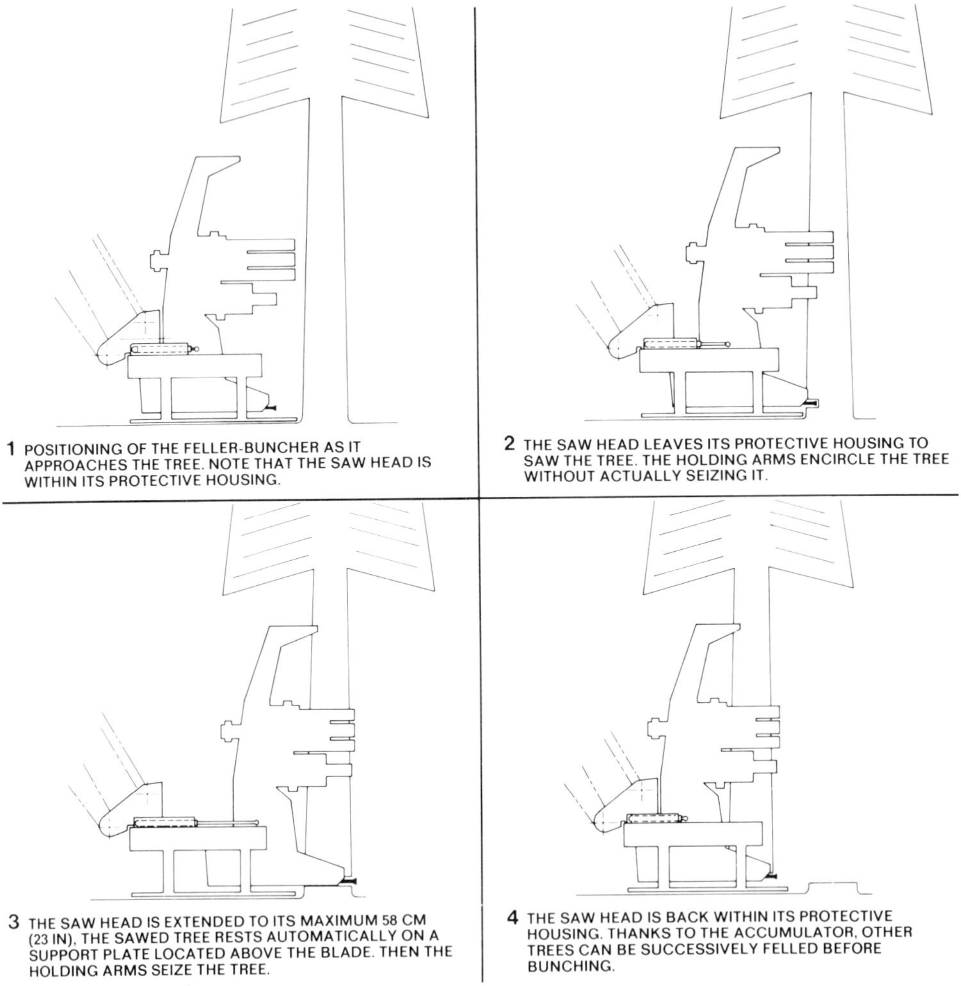

1 POSITIONING OF THE FELLER-BUNCHER AS IT APPROACHES THE TREE. NOTE THAT THE SAW HEAD IS WITHIN ITS PROTECTIVE HOUSING.

2 THE SAW HEAD LEAVES ITS PROTECTIVE HOUSING TO SAW THE TREE. THE HOLDING ARMS ENCIRCLE THE TREE WITHOUT ACTUALLY SEIZING IT.

3 THE SAW HEAD IS EXTENDED TO ITS MAXIMUM 58 CM (23 IN). THE SAWED TREE RESTS AUTOMATICALLY ON A SUPPORT PLATE LOCATED ABOVE THE BLADE. THEN THE HOLDING ARMS SEIZE THE TREE.

4 THE SAW HEAD IS BACK WITHIN ITS PROTECTIVE HOUSING. THANKS TO THE ACCUMULATOR, OTHER TREES CAN BE SUCCESSIVELY FELLED BEFORE BUNCHING.

FIGURE 34. Working principle of the circular saw feller-buncher.

The holding arms are synchronized with the sawing head at forward sawing, finishing the felling operation only when the tree is completely severed from the stump. This will prevent tensions with ensuing timber defects in the butt end of the trunk. A plate underneath the unit protects the saw from damage near rocks and similar obstacles. The circular saw blade, therefore, can be kept sharp longer.

Under snow conditions in winter the unit can be lowered to the bottom in order to reduce the stump height. The saw has a high capacity; at 1000 rpm a tree with a diameter of 50 cm at the stump will be cut off in the unbelievably short time of less than a second.

Several trees can be cut by means of the accumulator arm before laid down. A side tilt mechanism makes it possible to operate the base machine also in slopes and on uneven ground without affecting the felling saw.

The hydraulic sawing head can be controlled by the operator maneuvering a multiple control lever (Harricana Metal Inc, 1983).

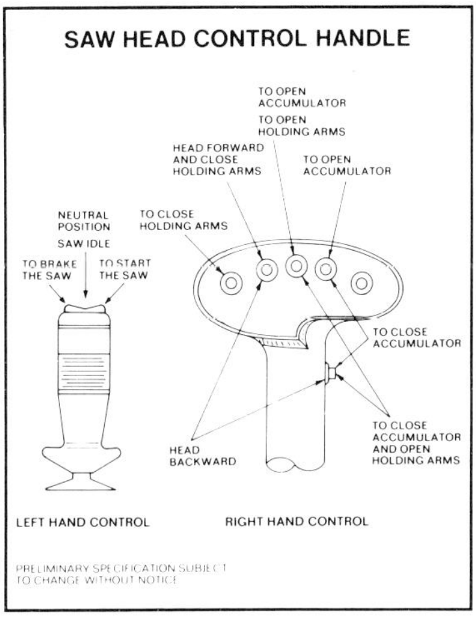

FIGURE 35. Saw head control handle

Felling head with two circular saws (manufactured in Quebec, Canada)

The felling head consists of two circular saws which are relatively small and thin. They produce a narrow kerf, eliminate timber defects in the butt end of the tree and they require a low power input. The circular saws proceed simultaneously through the trunk. The operation of the felling head is controlled by means of an on/off switch on the handle. The saw blades return into starting position in a protective housing immediately after cutting is finished (Equipment Denis Inc., 1983).

Technical data:				
	Height:	210.8 cm	2 engines	Vane type
	Width:	130.8 cm	2 infeeds	
	Length:	139.7 cm	2 infeed cyl.	2 "Ø x 30"
	2 saws	61.0 cm	2 clam cyl.	4" Ø x 7"
	Clam opening	81.3 cm		
	Total weight:	1452 kg		
	Working pressure:	176 kg/cm^2		

FIGURE 36. Felling head with two circular saws.

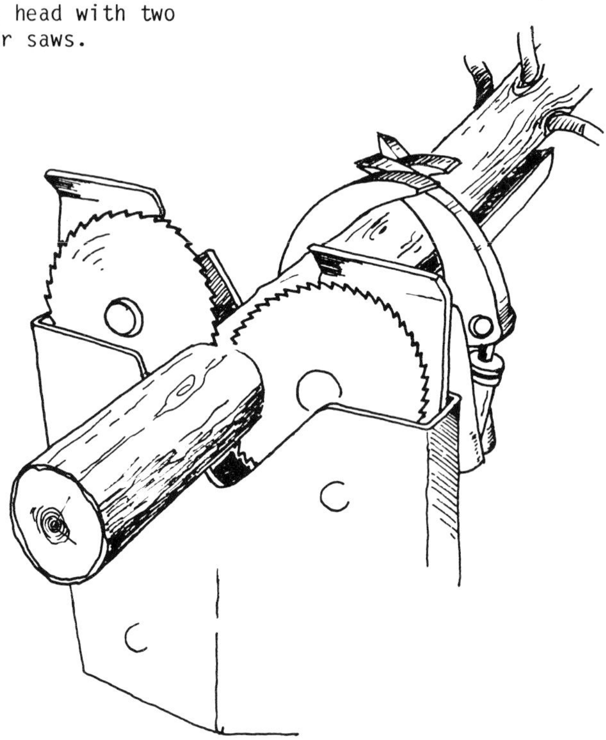

The horizontal surface of clipping tools, shearing tools and felling saws now existing on the market could be reduced to three times the cross-section area of the largest trees that can be felled if present development trends continue (Figure 37).

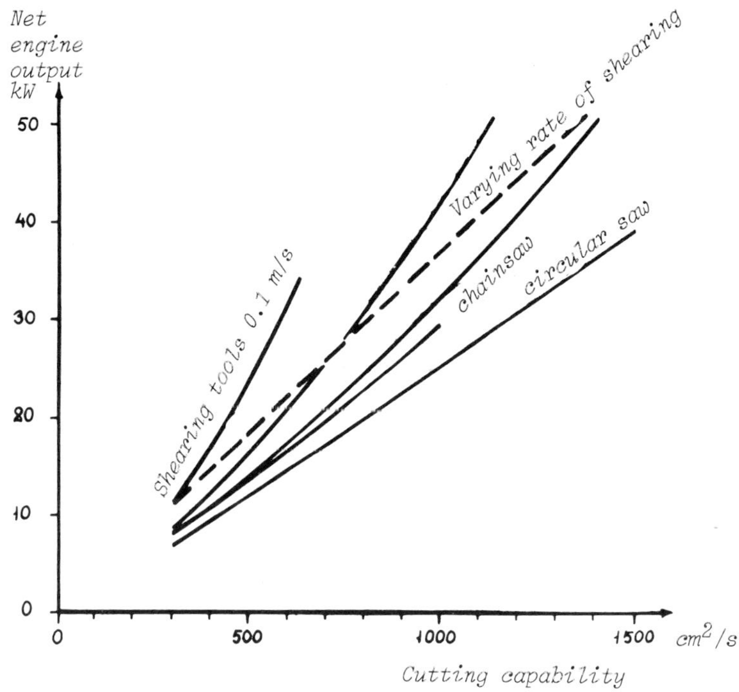

FIGURE 37. Relationships between the net engine output expressed in kW and cutting capability expressed in cm^2/s for various types of felling equipment (Wiklund, 1967).

Previously it has always been tried to cut as low stumps as possible in order to recover all useful wood. However, for practical reasons there is a limit. Felling becomes difficult and the butt swell of big trees is very cumbersome.

Possibilities to switch from conventional separation of the tree from the stump by means of a horizontal cut are being explored. Such a change would provide an opportunity to utilize those parts of the root system which can be recovered economically.

Investigations have shown that fibers in stumps and roots down to a diameter of 25 mm are good material for pulping.

Excluding roots smaller than 25 mm, volume of the root system is approximately 20 percent of the total tree volume.

Some minor investigations have shown that a large part of the stump, which is an extension of the trunk into the ground, can be used for lumber as well as pulpwood. Since it appears possible to recover this part of the stump together with the trunk without excessive cost, the method appears interesting.

The additional amount of timber that will be utilized by an extension of the butt log downward can be estimated. Separate experiments have shown that it is possible to lower the stump height by 20 cm. The value of this additional length of timber may be considerable.

According to current judgement, it appears reasonable to continue the experiments. Unless the investigations give negative results, the development of felling machines that facilitate a more complete utilization of the stumps will continue.

Alternative solutions

Several alternative technical solutions are probably required for further mechanization of felling operations, each solution designed for special conditions in order to produce the best possible result.

It is rather obvious that felling should be mechanized quickly as it is desirable for many reasons to eliminate or reduce the heavy and hazardous elements of work.

Felling patterns

The pattern of felling should be compatible with the methods of subsequent processing and transport.

Felling along strip roads

Directed felling is applied at conventional harvest operations e.g. felling along strip roads with bucking of timber into assortments (short timber) and piling of timber along the strip roads.

This pattern of felling facilitates transport of timber towards strip roads and to points along the roads where the timber is to be piled. Trees standing in or adjacent to the roads are felled at an oblique or straight angle to the roads depending on the length of the timber and on distance to the roads.

To facilitate subsequent processing, the trees are felled on top of each other in order to obtain a working height suitable for delimbing and bucking etc. A form of work bench is arranged (bench method).

Felling must take into consideration the number of assortments and the requirements concerning piling. It is also influenced by the leaning direction of the trees, terrain conditions, wind velocity, remaining stand and the occurrence of valuable natural regeneration. Proper direction of felling also facilitates piling of timber in the right direction of transport. Slash deposited in the tractor path is often valuable since it reduces rutting and improves the carrying capacity on soft ground (Figure 38).

Parallel felling and felling in swaths for the tree length trunk method

When felling is done by means of chain saws for transport of trunks according to the tree length method, two patterns are used viz. parallel felling and felling in swaths (Figure 39).

Parallel felling is adapted to skidding of trunks when the butt ends are lifted either by means of choker cables or by clam bunks. The trees are felled parallel to each other, hence the term.

Felling in swaths is adapted to skidding of trunks when the top ends are lifted by means of choker cables. Felling in swaths without strip roads is preferable if a single drum winch is being used. The top ends of the trunks are then easily collected if felling is directed along swaths.

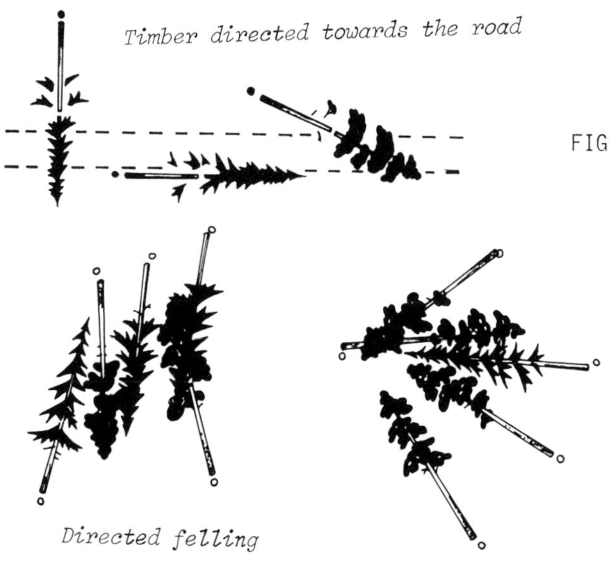

FIGURE 38. Slash in the strip road improves the carrying capacity of the ground and rutting is reduced.

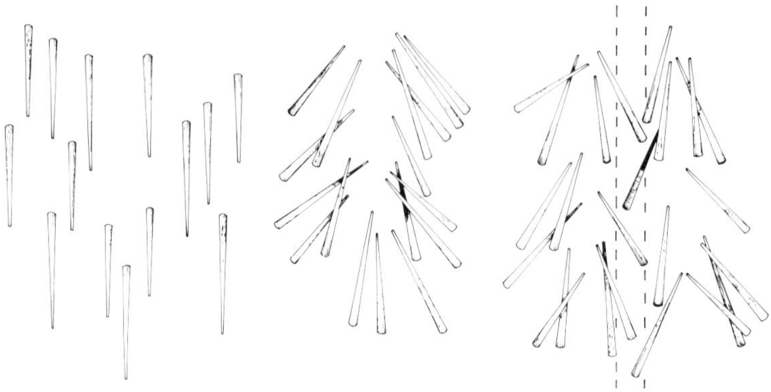

FIGURE 39. a) Parallel felling adapted to skidding with butt ends lifted by means of choker cables (couplings) or clam bunks.
b) felling in swaths without strip roads adapted to skidding when top ends are strapped or choked and lifted by means of a single drum winch.
c) felling in swaths with strip roads adapted to skidding when top ends are strapped or choked and lifted by means of a double drum winch.

Felling in swaths with strip roads is preferred when skidding is done by means of a double drum winch. The top ends of the trunks are then collected in two swaths, one swath on each side of the strip road, from which the trunks can be winched simultaneously.

When the tree length method is being used, a well arranged directed felling is of great importance for a quick loading operation. A poorly arranged directed felling may increase the time required for coupling or choking by up to 15-30 percent when the top ends are pulled first.

The angle between the road and the trunks must be adjusted primarily with respect to boulders in order to prevent that the trunks jam the winch. A wide angle between the road and the trunks gives a larger amount of timber in the swath than that obtained when the angle is narrow.

The biggest trees should be felled first in order to provide for a more convenient bunching of the small trees which are felled on top of the big trees. This arrangement will also facilitate delimbing.

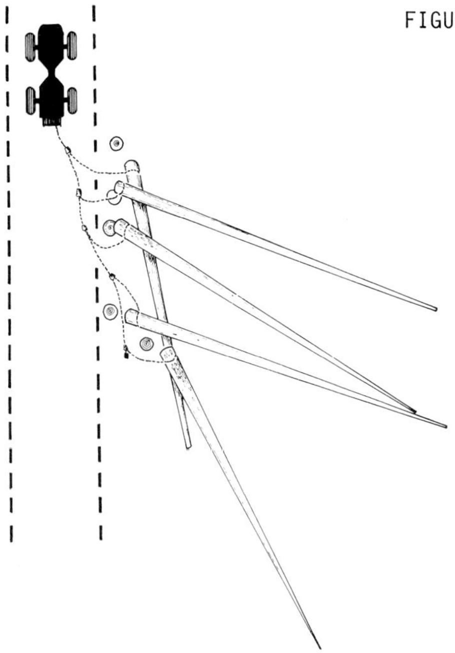

FIGURE 40. Felling on top of another tree in order to facilitate coupling or choking of butt ends.

When transport is done with butt end first, felling should be directed over another tree in order to facilitate coupling (choking) and winching of the trunk to the tractor (Figure 40).

Parallel felling for the tree method

When the tree method is being used, parallel felling is usually applied for transport of trees with the butt end first, skidding by means of clam bunk often being the method used.

Delimbing and topping before felling

When the harvesting operation is carried out by means of e.g. a delimber-feller-buncher, the tree can be delimbed and topped in standing position before it is cut off at stump by means of a clipping or sawing mechanism and laid down into bunches of two or several trunks. This form of felling is also applied by manual methods when trees in parks or other places cannot be felled directly. The trees are then cut into logs from the top down to the ground. Certainly, this manual method would be both expensive and risky if applied in the forests.

Extraction of trees in vertical position

A "felling" method has been described according to which the tree is lifted after being separated from the stump and moved laterally in vertical position to a place for processing.

At mechanized delimbing and other processing the direction of work movements is often determined at the bunching of trees in horizontal position to the processing machine. This applies in particular to stands where some trees remain after the harvesting operations e.g. in thinned stands. If the cut trees can be moved in a vertical position to the processing machine or the tractor standing on a strip road, there are several possibilities for the subsequent processing and handling of the tree from above, from the front or from the rear through the machine when it is moving along the strip road.

Of special interest at this form of "felling" and bunching of trees are the lifting leverage and the turning and bending forces which will be required because of friction between the crowns of the cut trees and the crown canopy of the remaining stand.

Technical data necessary for the construction of booms feasible for this form of felling and transport have been obtained from observations of the forces in a test bench (Myhrman, 1968).

Data of particular value for manufacturers of e.g. cranes were obtained at measurements of crowns of pine and spruce trees weighing between 150 kg and 600 kg. Absolute and relative crown contact was studied by varying the distance between the tree braces.

Tests have shown that an acceleration of 5 m/s^2 is a suitable value giving short acceleration and deceleration times and a reasonable stress on the boom.

The maximum bending leverage for a 30 cm (Dbh) tree with a weight of 650 kg may amount to 36 000 Nm, and the maximum turning leverage on the boom to 5 000 Nm. No significant difference has been found between measurements obtained in winter and summer conditions. Damages to the crowns of 'remaining' trees appeared slight.

Lifting trees for processing in a vertical position is so far subject to theoretical analyses only. Field experiments have been carried out by means of a so-called tower crane equipped with a felling device for certain tests of positioning.

Lifting the tree vertically from the stump provides several interesting aspects on the problems of tree extraction in thinning. It would avoid the problems of crowding and crown friction. A considerably reduced processing cycle per tree would be achieved because processing could be initiated immediately after the tree has been severed at the ground level. However, the method is fraught with problems of positioning and visibility.

Test results indicate that at least three trees per minute (small trees from thinning) must be harvested from above in order to make the method viable. The clipping - sawing mechanism is then moving the shortest possible way between the trees.

Collection of trees could be done according to alternative 3 in Figure 41.

Alternative 1: Trees are felled for collection in horizontal position at the strip road

Alternative 2: Trees are lifted in vertical position for collection at the strip road.

Alternative 3: Trees are lifted vertically for processing and transport above the crown canopy to the strip road.

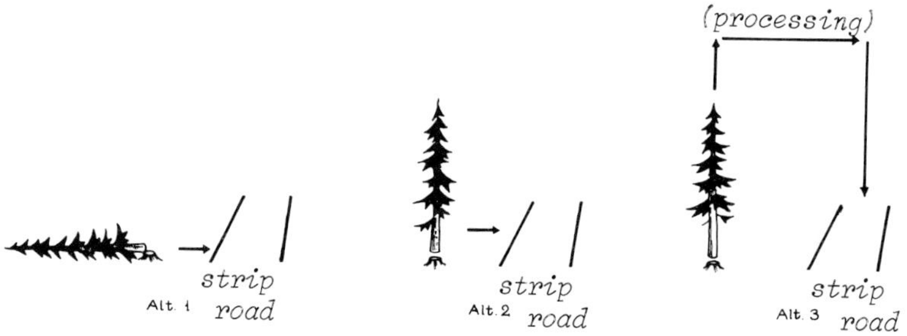

FIGURE 41. Various alternative transfers of trees from the stumps to the strip road.

Felling of whole trees

Felling of whole trees with their entire root systems is an old method used by our settling ancestors. When the tree trunk was used as a lever, the roots could be extracted from the ground. Felling could be facilitated by cutting off root branches as far out from the trunk as possible. A similar type of felling occurs at strong winds.

Lifting of whole trees

Trees can also be lifted with their entire root systems. This method is worth considering in particular when the supply of timber is short. The additional volume of useful timber in the central core of the root system is estimated to be 20 percent of the useful timber. The method is subject to experimentation in several countries and for various special situations, e.g. in Brazil, Canada and Finland.

Trees felled with cut root systems

A tree is felled rather easily if the root system is cut.

A variety of this method is to lift the tree after the root system is cut. Machine equipment can be used to cut off the roots straight down along the tree trunk or at an oblique angle towards the trunk in order to recover the valuable extension of the trunk into the stump. The method is being developed in several countries.

Felling or collection of several trees simultaneously

In stands where thinning is being carried out, a knuckle boom equipped with a clipping or shearing mechanism or a felling saw with a basket for a simultaneous collection of several trees appears to provide a potential solution.

To reduce the processing cycle for trees removed in thinning, felling mechanisms for collection of two or more trees simultaneously (accumulators) are currently being developed for stands with very small trees.

Experiments with a recently developed felling mechanism have shown that the processing time per tree can be reduced considerably when 2 or 3 trees are collected simultaneously (Bredberg, Moberg, 1972).

FIGURE 42. Feller designed for simultaneous handling of several trees.

FIGURE 43. Feller equipped with conic circular saw which can fell simultaneously 2-3 trees standing adjacent to each other (ÖSA, 1982).

Manual felling

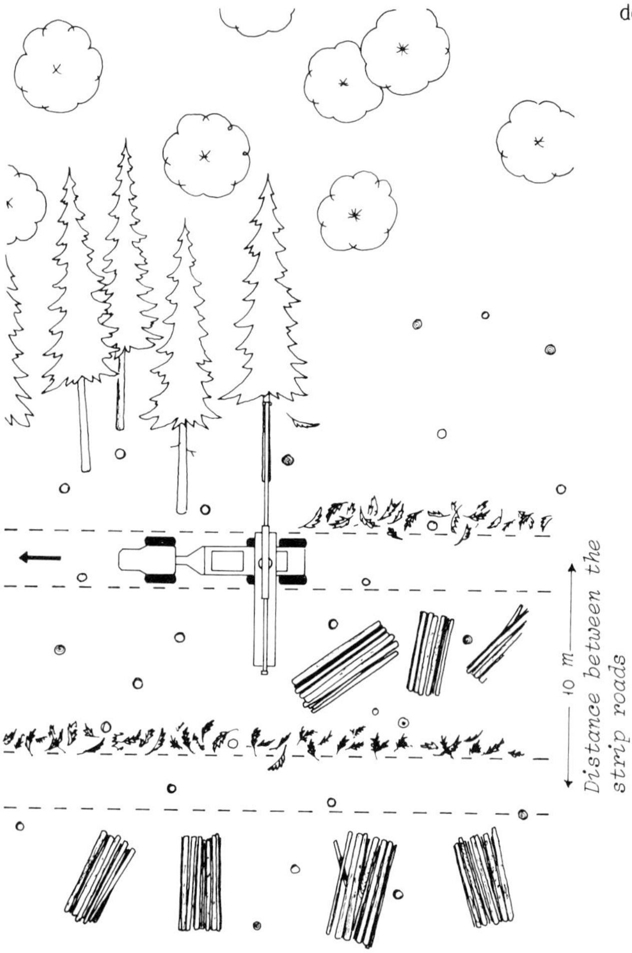

FIGURE 44. Felling for mechanized delimbing and bucking.

Manual felling with mechanized processing

Manual felling in combination with mechanized processing can be carried out by means of the following types of machines

- Telescope delimber starting from the top of the tree
- Telescope delimber in combination with a bucking saw for pulpwood

- Telescope delimber with bucking saw
- Delimber with bucking saw equipped with fixed tools.

Differences that occur concerning the felling patterns when the machines listed above are being used are not judged to be influential on the situation with respect to time studies. Time required can be calculated according to the same formula in all cases of felling.

Work is generally carried out in the following way: The feller is equipped with a chain saw and a lever. Directed felling is very important. Felling for a delimber - bucking saw with fixed delimbing mechanism is assumed to be arranged by first making a corridor through the stand. Felling is then done in a general direction towards the corridor. In other cases the trees can be felled in a direction away from the border of the stand (Figure 44).

Alternative tree part method in thinning operations using grapple saw on crane with long boom

Equipment:

1. Crane with long boom reaching 11 m
2. Crane has a grapple saw with counter supports in both sides
3. The saw is suspended in a rotator with hydraulically controlled links which can lift the timber without swaying
4. Crane and grapple with electric-hydraulic power operation and pedals
5. Chain saw for motor-manual felling according to a predetermined pattern
6. Forwarder

The pattern of felling when the tree part method is used in thinning operations may vary. The figure shows a pattern suitable for the equipment listed above.

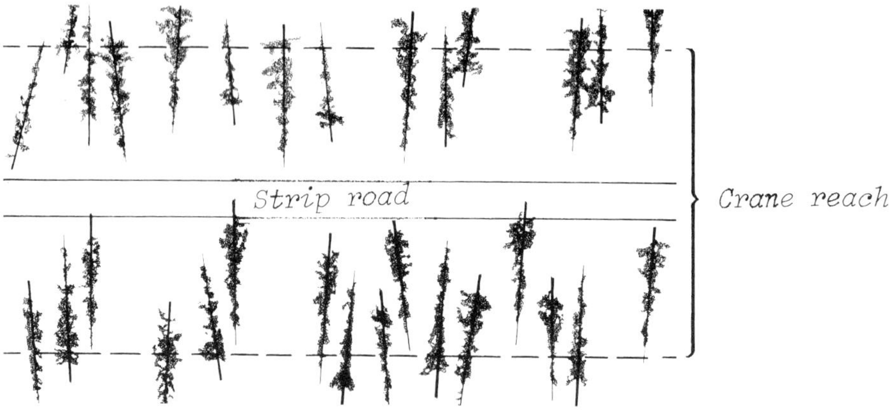

FIGURE 45. Felling pattern when the tree part method is used in thinning operations. Trees standing within the crane reach are felled away from the strip road while trees that stand outside the crane reach are felled toward the strip road.

Comparisons between the various tree part methods and assortment methods have shown that the tree part methods with a grapple saw mounted on a crane with long boom can be an alternative worth further developing and testing (Österlöf, 1981).

FIGURE 46. Harvester in operation.

Mechanized felling

Five different forms of mechanized felling are compared in the following presentation.

At this point it may be mentioned that a machine that is capable of felling trees can also be called a harvester while a machine that is not capable of felling trees is called a processor (e.g. bucker-delimber-buncher).

Feller - a small skidder with straight boom

Description of machine and method

The machine consists of a small skidder equipped with a crane, in practice reaching 5 m out. A felling device, mounted at the head of the boom, is designed to prevent the forces exerted by the tree from being absorbed by the machine. Felling can be directed (Figure 47).

This method of felling replaces the manual felling. When felling is done for machines that process the trees from the top, the trees are felled towards strip roads which are 20 m apart.

When felling is done for machines that process the trees from the stump, the trees are instead felled parallel to and straight backward in relation

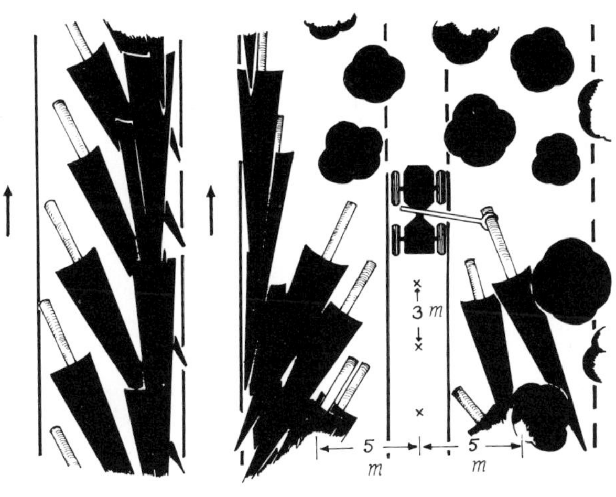

FIGURE 47. Felling parallel to the roads for mechanized processing from the top of the trees.

to the feller advance. Travel speed, positioning time and other time data can be obtained by time studies, from statistics or after theoretical calculations.

The machine should not fell the trees into the stand in front of it. It is necessary, therefore, to anticipate a time allowance of 10 - 20 cmin per tree for idle drive.

Feller mounted on a tracked vehicle with short boom

Description of machine and method

The machine consists of a small, approximately 2 m wide base machine with a felling mechanism mounted on a 3 m long boom. Engine output is approximately 22 kW. Prominent features of the machine are low weight (approx. 1 tonne) and good terrain travelling ability. Since the crane is designed only to bring the felling mechanism to the tree, the machine cannot handle the tree to any great extent.

Felling can be directed and the machine is designed to prevent trees from falling backwards.

The idea behind this design was to retain the advantages of the method "Man with chain saw", the difference being that the operator is given a considerably improved working place in a safe cabin.

The machine proceeds between the trees along a zone, felling the trees directed either towards the strip roads 20 m apart (processing from the tops of the trees) or at a straight angle to the direction of the machine movement (processing from the butt ends of the trees). Felling zone is calculated to be 4 m wide (Figure 48).

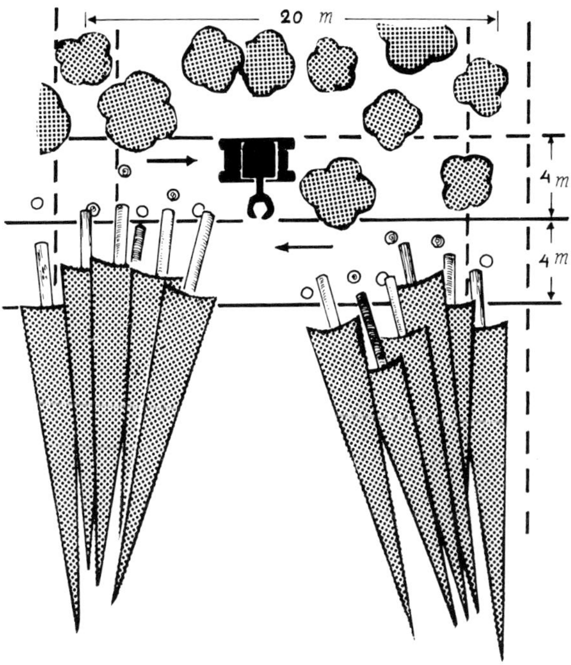

FIGURE 48. Felling toward strip road for mechanized processing from the tops of the trees.

Feller

Description of machine and method

A stable base machine with good terrain travelling ability e.g. a big skidder or excavator, is equipped with a felling mechanism mounted on a straight boom. The machine proceeds in the stand separating the trees from the stumps, lifting and turning the trees and laying them down on the ground in a desired direction. Laying the trees in a string or a certain amount of bunching of trees that can be reached from the same position can be done. However, the machine cannot transport the trees (Figure 49).

The practical reach of the machine is considered to be 6 m which gives a swath of approximately 12 m. Distance between the positions is normally assumed to be 3 m.

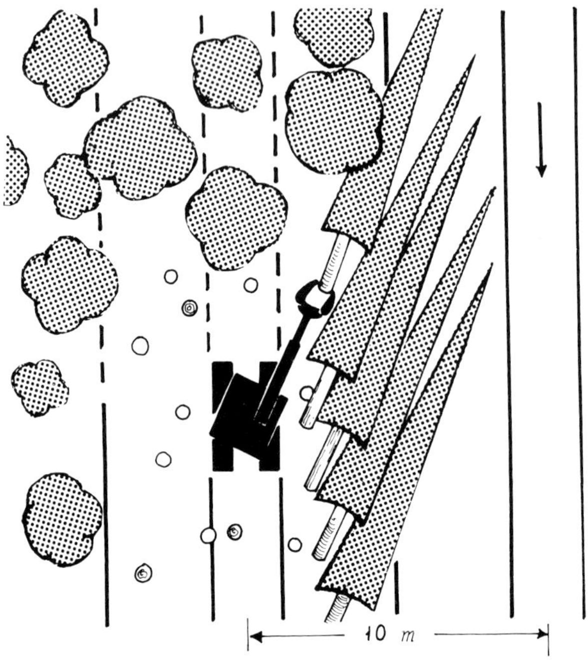

FIGURE 49. Mechanized felling of trees.

Feller - buncher

Felling - bunching is done in combination with the following work operations: skidding by means of clam bunk, telescoping delimber starting from the stump and with a flexible choice of position for the telescopic delimbing - bucking saw.

Description of machine and method

The machine can be operated according to two main principles either as a feller - buncher laying bunches at strip roads, or as a feller - skidder without reloading, transporting the trees to landings at truck road.

The base machine is a forwarder with an engine output of approximately 110 kW. Felling is done by means of a mechanism mounted on a knuckle boom having a reach of approximately 6 m from pivot of the crane. In the rear of the machine is placed a clam bunk with a loader of approximately 2 m^2 cross-section capacity. Size of load varies with the method of work, terrain difficulties and tree sizes.

The shortest distance between the positions is assumed to be 3 m. Working on one side only, the machine may cover a swath of 5 m (Figure 50).

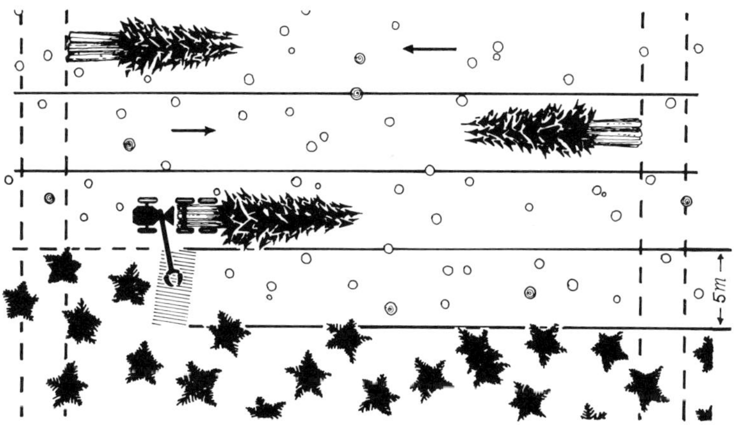

FIGURE 50. Mechanized felling and bunching or direct skidding.

Feller - skidder (buncher)

Description of machine and method

The machine is similar to the feller-buncher described above. The only difference is that the feller-skidder transports trees to a landing instead of leaving bunches in the stand.

Feller - delimber - buncher (felling integrated with other operations)

Description of machine and method

The base machine is a six-wheel drive forwarder with a hydraulic steering of the frame, an engine output of approximately 110 kW and a hydrodynamic gearbox. The felling-delimbing unit, the bunching mechanism and the driver's cabin are placed on a cog ring above the bogie centre. The combined felling and delimbing mechanism is supported by the base of a two-part telescoping crane (7 m + 7 m). Felling is done by means of a chain saw for a maximum tree diameter of approximately 50 cm.

FIGURE 51. Mechanized felling, delimbing and bunching by means of one machine.

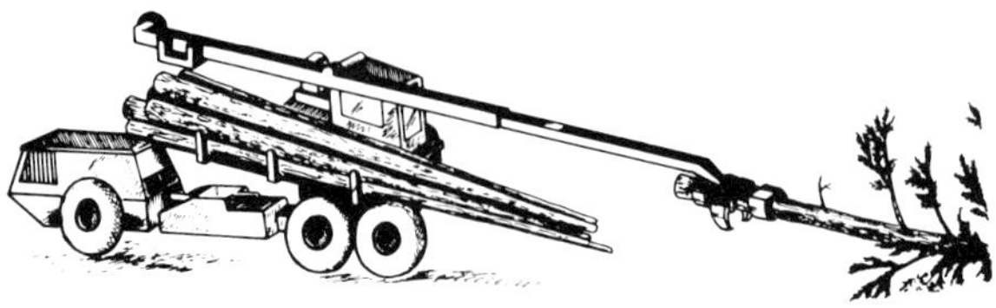

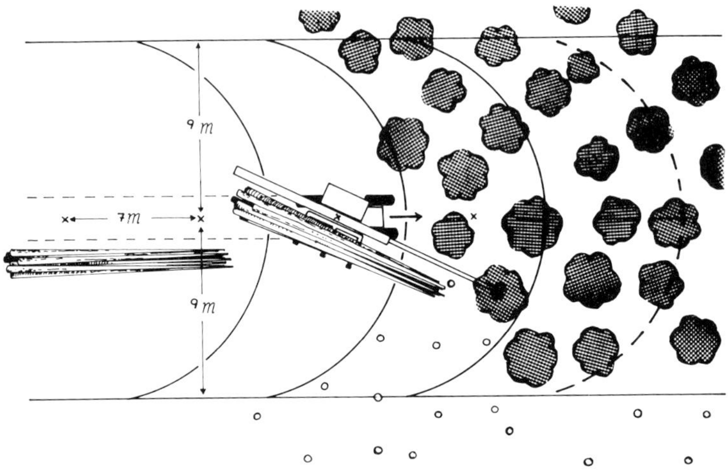

After severance, the tree is felled forward while being pulled in towards the machine. When the telescoping boom is in its innermost position, the butt end is grappled by a carrier moving along the posterior fixed boom. The carrier moves backwards while the telescoping boom moves forward. Delimbing is carried out by means of trunk embracing knives. Topping is automatic at a predetermined minimum diameter. Reach of the crane is 9 m from the pivot (Figure 51).

After delimbing and topping, the trunk is automatically transferred into a cradle which is emptied when a bunch of given size has been accumulated. Size of the bunch varies with tree sizes.

Several examples of felling operations integrated with other harvesting operations are shown in subsequent chapters.

Feller-delimber-bucker

The felling mechanisms developed in the 1960's and the grapple equipped harvesters, i.e. units for combined felling and bucking of the 1970's have been combined in the crane mounted grapple harvesters of the 1980's. The new modern machines carry out all the partial operations such as felling, delimbing, bucking-scaling and bunching.

The design of the grapple harvester was conceived relatively recently and most of the types developed, therefore, may be considered to be prototypes at the testing level (Figure 52).

At final harvest scaling is a problem on account of unsafe stopping of the tree when feeding is done by means of spike rollers or rubber covered rollers. The cranes are also subject to great stress because of the heavy felling units (heads).

Grapple (clam) harvesters will primarily be useful at the mechanization of thinning operations. The conventional harvesting machines must still be used for final harvest operations in heavy and coarse-limbed timber. Chain saws for felling, delimbing and bucking and forwarders will remain the most common machines for harvesting of very large trees.

FIGURE 52. A feller-delimber-buncher.

Some performance data

A comparison of the performance of the manual and mechanized forms of felling presented above is reported here for final harvest operations typical of conditions in the northern boreal region (Table 4).

Performance varies strongly among the various forms of felling in the same type of stand, in the table below between 39 trees and 143 trees per efficient working hour. Variation depends on i.a. pattern of felling, combination with other work operations other than pure felling such as bunching and transport. Costs of the various forms of felling become of interest only when they are included in complete systems of harvesting.

Table 4. Some performance data for manual and mechanized forms of felling in stands typical of conditions in the northern boreal region (Nilsson, 1968).

Form of felling	Cmin per tree	No. trees per hour	No. trees per shift	m^3/shift
Manual felling				
chain saw	137.5	43	258	77
Mechanized felling				
Feller-small skidder with straight boom	42.0	143	715	215
Feller-scooter with short straight boom	47.7	125	625	188
Feller	52.2	96	480	144
Feller - buncher	79.4	76	380	114
Feller - skidder	154.7	39	195	77
Feller-delimber-buncher	47.5	126	630	189

Un-manned machine without operator seat in the cabin

A new type of machine is a mini-skidder rebuilt into a feller-skidder with a supplementary felling head for thinning in swaths. It is designed for handling bunches of small trees. Since the machine enters into the swaths once only, the distance between the strip roads must be adjusted to the volume of timber removed in each swath. Distance between the strip roads is normally 40 m - 60 m and distance between the swaths is 4 m - 6m (Alriksson, 1983).

The operator maneuvers the machine from the ground by means of a steering stick for moving and a lever for controlling the crane at the front of the engine for felling and bunching.

Operation is based on a number of hydraulic engines coupled in series and parallel to each other providing all-wheel drive.

On soft ground the strip roads are cleared and covered with residues. On firm ground the feller-skidder itself is harvesting the trees in the strip roads.

The machine enters the swath by backing into it, felling the trees in the way. The trees are laid to the sides and loaded at the exit from the stand. On its way out of the swath the machine is thinning the stand on both sides and the trees are laid on a clam bunk with 0.8 m^2 cross-section area. The load capacity is approximately 1.0 m^3 - 1.5 m^3 (solid wood), corresponding to approximately 25-50 trees.

In its application this machine has the following advantages:

- slight damages to ground and the trunks of remaining trees
- reduced risks of windfelling due to the small area of strip roads
- good thinning effect because of individual selection of trees from the ground
- higher volume of timber removed per hectare
- optional handling in the further transport
- high concentration of timber at the strip roads
- timber evenly bunched, facilitating the use of grapple saw forwarding
- clean timber, free of soil and rocks
- machine can travel on soft ground
- simplified planning of operation
- low cost of moving between projects
- high degree of technical utilization

For efficient operation the machine requires:

- uniform objects without difficult undergrowth
- unsorted bundles of various species
- daylight, light snow cover (max. 0.3 m) and trees free of snow

Output of the machine is approximately 3 m^3 (solid wood)/h

Technical data:	Width	160 cm
	Weight	1 850 kg
	Clearance	32 cm
	Engine, 2-cyl. air cooled, diesel	18 kW
	Transmission, 2 circuits	2 x 14 l/min
	Hauling capability	2 000 kp
	Crane, reach laterally	2.9 m

Shearing diameter, felling head	25 cm
Price excl. tax (1983)	230 000 SEK (1 $ U.S. = 8 SEK)

FIGURE 53. Shearing-feller-skidder in operation.

Small machine for felling and bunching in thinning operations

In the first years of the 1980's several models of small tractors with tracks or wheels have been developed for first and second thinning.

Example. Machine with 8 wheels for felling and bunching. It operates in the stand in high piling and even bunching of the trees for a high capacity in subsequent handling or processing (Alriksson, 1982).

FIGURE 54. Machine for felling and bunching in thinning operations. The machine is only 176 cm wide at base. Tapering upward, it has a low centre of gravity, good stability and less contact with the tree trunks (less damages to the remaining trees).

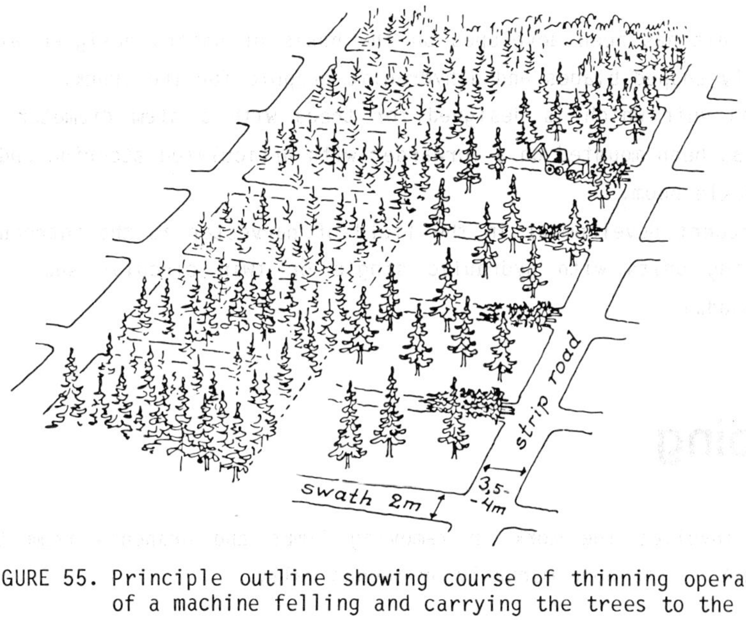

FIGURE 55. Principle outline showing course of thinning operation by means of a machine felling and carrying the trees to the strip roads.

All movements in the stand are along swaths perpendicularly to the strip roads in order to minimize length of the swaths and the hauling time of trees.

elopment of hydraulic mechanisms such as felling clippers ..g saws, the felling machine (harvester) has now definitely been introduced in forest operations. The feller-buncher has become the dominating machine and a large number of units are employed in practical operations.

If we assume that each machine is felling an average of one tree per minute, or 50 trees per efficient working hour, and if it is used 1000 hours annually, the output of the machine per year will be 50 000 trees. If 100 machines are operated in final felling operations, performance will be 5 million trees annually. This is mentioned in order to give an idea of the magnitude of operation and the extent that can be expected in future felling operations. Mechanization of felling and bunching eliminates two very heavy manual work phases.

The felling mechanism, which is built for trees with a maximum stump diameter of 50 cm, has been mounted directly on a tractor or on a forwarder with articulated steering and a hydraulically operated knuckle boom.

A felling unit has been developed on the basis of shears designed with two spherically shaped blades and a hydraulic grapple for the trunk.

The felling unit which is designed for trees with a stem diameter of about 50 cm has been mounted on a forwarder with articulated steering and a hydraulic knuckle boom.

The most recent development in the field of harvester is the introduction of felling units with hydraulic single or twin circular saws (in U.S.A. and Canada).

Delimbing

Delimbing involves the work in removing limbs and branches from the tree. This work can be done manually or mechanically.

Manual delimbing

Approximately half the working time required for the processing of a tree is spent on delimbing at conventional harvesting according to the assortment ("short wood") method.

Delimbing by means of an axe is decreasing steadily. Manual delimbing by means of chain saws is now a dominating work phase, both physiologically and in terms of cost.

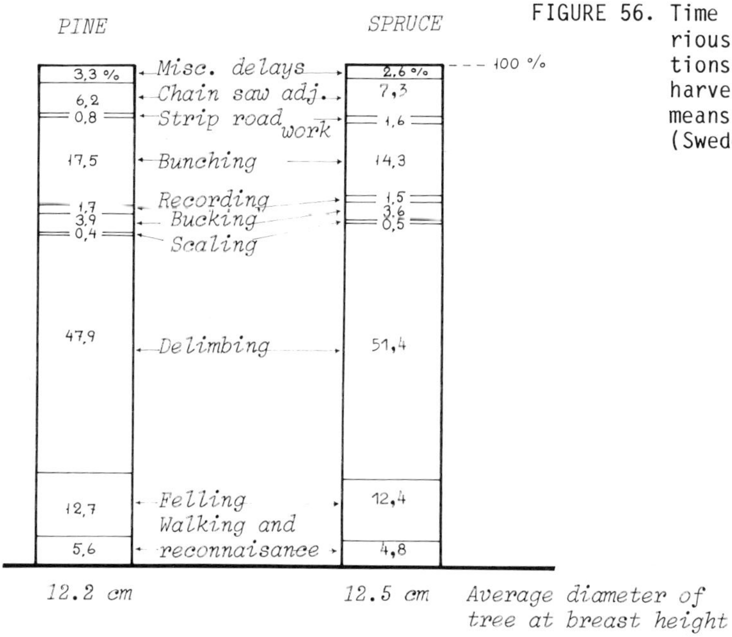

FIGURE 56. Time required for various partial operations at conventional harvesting of trees by means of chain saws (Sweden).

Figure 56 shows the proportions of time required for various operations when the assortment (short wood) method is applied to small trees. The distribution of time applies to processing of rough timber (no debarking) into random length in combination with directed felling, scaling without measuring device, delimbing by means of chain saw and bunching to strip roads.

Mechanized delimbing

Technical research and development work on the mechanization of delimbing has brought about a rapidly increasing use of delimbing machines in forest operations. Although mechanized delimbing is not yet fully developed from the point of efficiency, delimbing can now be carried out at a lower cost due to increased integration with other harvesting operations such as bucking and bunching.

rs mechanized delimbing has increased primarily in final s. The greatest gains have been achieved in very coarse-nds.

The tree limb as an object of work

An exact definition of branches and limbs does not appear necessary in this context. However, it might be of some interest to explore how the limbs are normally connected to the tree trunk, their characteristics, weight etc.

Whorls and internodes

Most of the limbs of coniferous species are placed in whorls. This is of some interest from the point of mechanization.

Distance between the whorls, which normally consist of 4-6 limbs, varies with height and age of the tree. This distance is called internode. Fast growing trees have longer internodes (> 50 cm) than have slow growing short trees (< 10 cm). Since increment in height declines with increasing age, the internodes in old trees are shorter and the diameter of the limbs decreases towards the top of the trees.

For a calculation of the forces needed for delimbing it is necessary to know the length of the internodes, number of limbs per whorl, frequency of limbs (no/m of trunk length), diameter of limbs, angles of the limbs, occurrence of green and dry limbs, and their resistance to shearing force. It is also of value to know crown length and weight of limbs.

Frequency of limbs

Number of limbs in the second metre from the ground, in the green crown is shown principally in Figure 57.

Diameter of limbs

Diameter of the limbs in the second metre of the green crown at various average diameter and height of trees is shown principally in Figure 58.

Diameter of the limbs has been measured at a point 5 cm from the surface of the trunk and parallel to the trunk. The average diameter of limbs from the whole green crown does not differ essentially from the average diameter of limbs in the second metre (from the ground) in the green crown.

At the surface of the trunks, diameter of the limbs is approximately 20-25 percent larger than at 5 cm from the trunk.

Height to crown base

Figure 59 shows an example of the height to the first dry limb, to the first green limb of the crowns, and total tree height for pine.

Resistance to shearing force at delimbing by means of cutting tools

Shearing force required for delimbing has been studied at laboratory experiments. The following formula can be used for summary calculations:

Pine: $F = 3\ D^2 + 75\ D$

Spruce: $F = 5\ D^2 + 100\ D$

F = shearing force expressed in Newton

D = diameter of limb in mm

Shearing force as a function of limb diameter measured perpendicularly to the direction of shearing at 5 cm from the surface of the tree trunk is shown as an example in Figure 60.

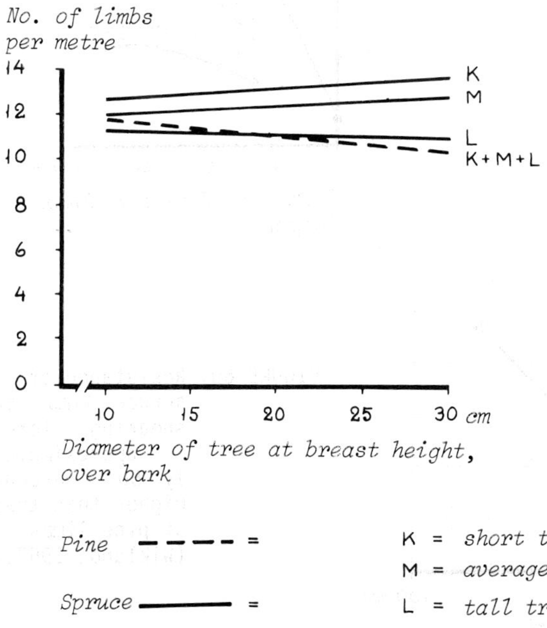

FIGURE 57. Number of limbs in second metre of the green crown for trees of various sizes (Ager, 1972).

FIGURE 58. Diameter of limbs in the second metre of the green crown for trees of various sizes (Ager, 1972).

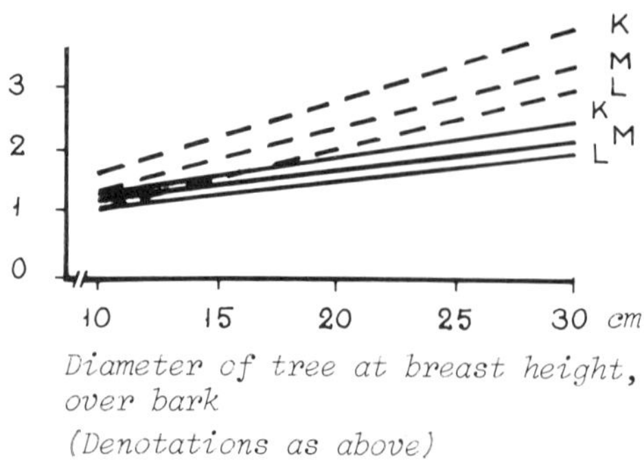

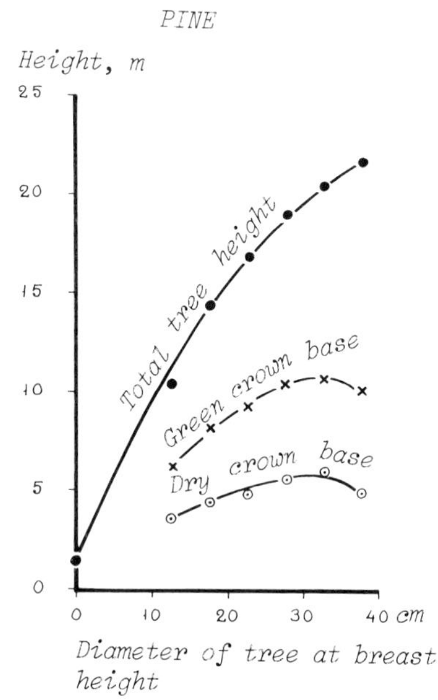

FIGURE 59. Example showing total tree height and height to crown base (first dry limb and first green limb). Pine.

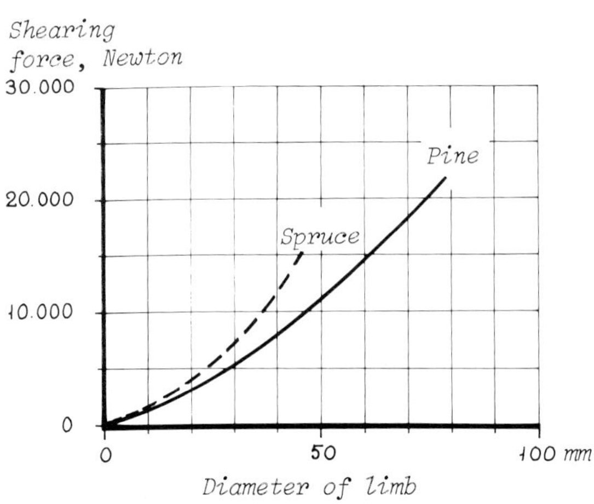

FIGURE 60. Resistance of spruce limbs to shearing force is approximately 50 percent higher than that of pine limbs (Wiklund, 1967).

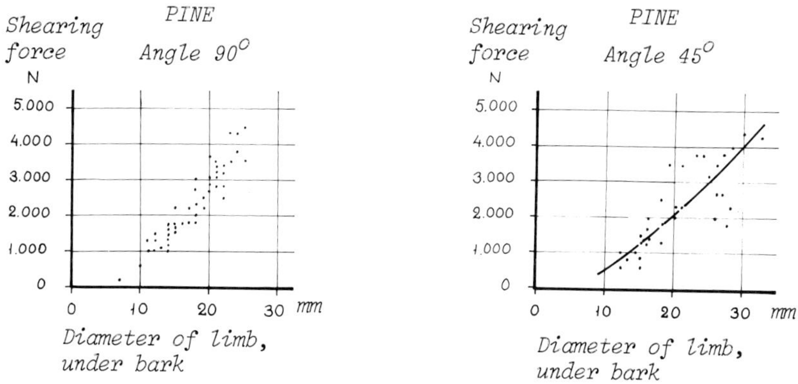

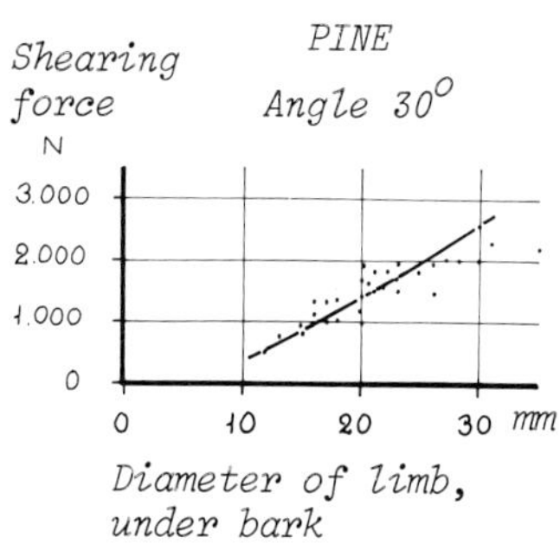

FIGURE 61. Example showing resistance to shearing force in pine limbs at various angles of the cutting edge (Callin and Forslund, 1968).

The graphs apply to fresh limbs of pine and spruce delimbed from the butt end of the logs by means of a 10.4 mm thick knife.

Some values of shearing force for limbs of small trees have been given as examples for various angles of the cutting edge (Figure 61).

Weight of limbs

According to weight measurements of trees and trunks a normally developed crown of spruce in Sweden weighs approximately 50 percent and 30 percent of the weight of the trunk in trees which are 8 cm and 20 cm (over bark) at breast height, respectively. Corresponding values for crowns of pine amount to 40 percent and 20 percent, respectively (Figure 62).

In young stands (thinning stage) the work objects (trees) are smaller and considerably lighter than in final harvest operations, the limbs are smaller, sounder and probably less densely placed. These are factors of importance for the development of machines for thinning and delimbing.

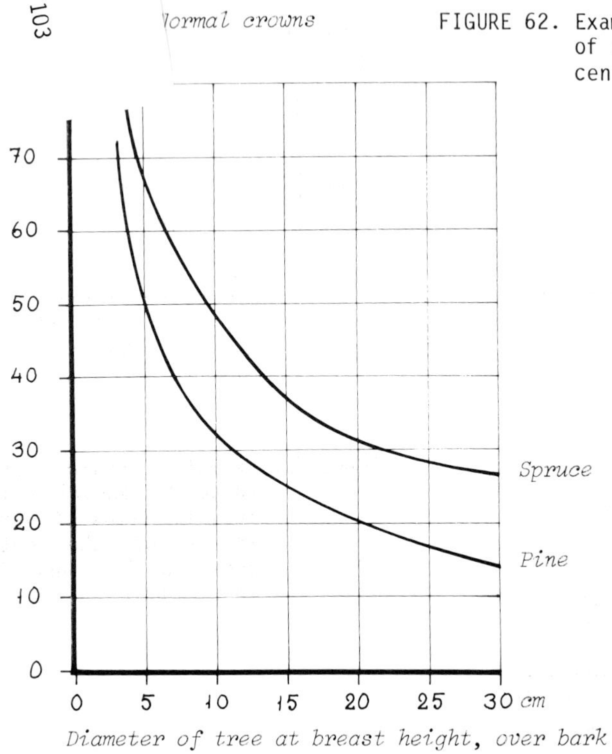

FIGURE 62. Example showing weight of normal crowns in percent of weight of trunks.

Tools and means of delimbing

Delimbing can be carried out at some of the following five levels of mechanization:

1. By means of hand tools: axe, spud, knife, saw
2. By means of motor powered hand tools, e.g. saws
3. By means of machines for delimbing only
4. By means of machines for delimbing integrated with other operations e.g. combination of delimbing and bucking
5. By means of machine systems with remote control. Some degree of automation is applied at e.g. industrial depots.

Various tools and machines for delimbing

1. Tools with edges

1.1 Trunk embracing knives and stepwise feed

1.2 Trunk embracing knives and roller feed

1.3 Trunk embracing knife track and stepwise feed

1.4 Trunk embracing knife track and roller feed

2. Tools with cutters

2.1 Fixed cutters and roller feed

2.2 Moving cutters and roller feed

3. Tools with flails

4. Tools with screws

5. Tools with chains

5.1 Fixed chains

5.2 Rotating chains

A presentation of some machine types

The following machines are examples of various technical solutions and methods available for practical use.

Trunk embracing knives and stepwise feed

Delimber - buncher

Type of delimbing tools: Trunk embracing knives, one fixed and two moving, mounted on a telescopic boom (Figure 63).

Removal of slash (limbs)

Slash is scattered along the 7 m straight boom, tops are cut off and deposited in front of the machine.

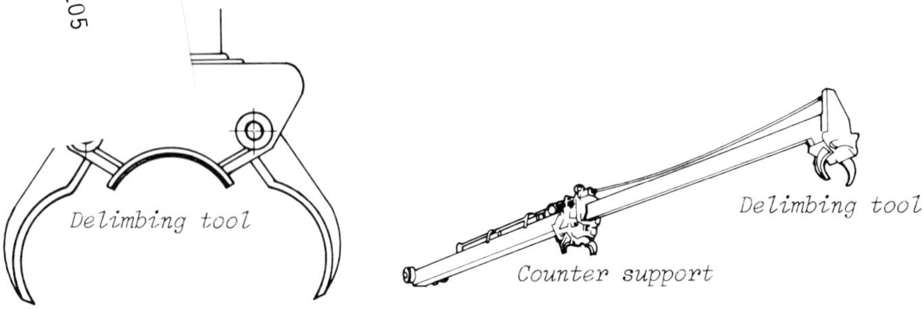

FIGURE 63. Trunk embracing knives, one fixed and two moving, mounted on a telescopic boom with counter support.

Tree harvester

Type of delimbing tools: Trunk embracing knives, one fixed and two moving.

Infeed: The delimbing tool is attached around the standing tree (Bjerkelund, 1965).

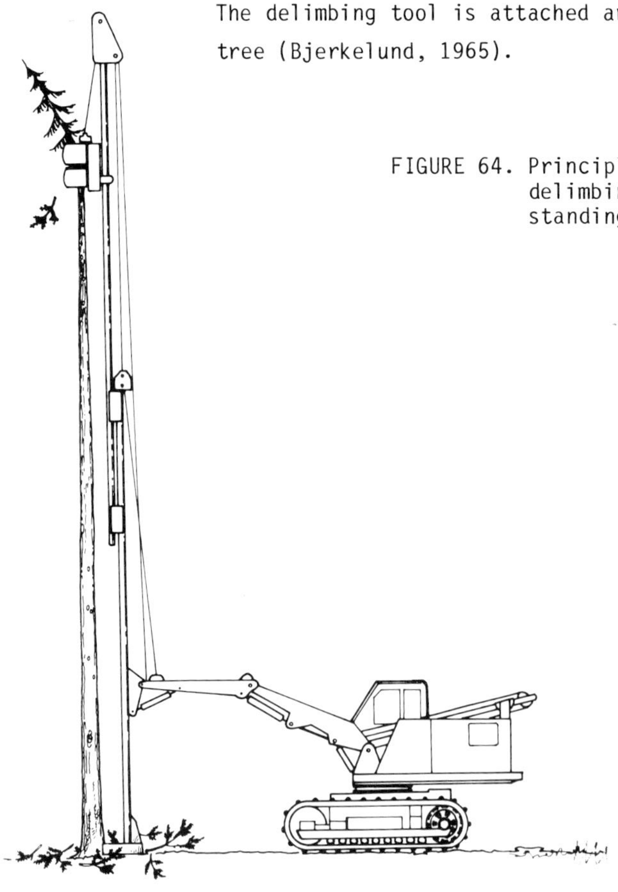

FIGURE 64. Principle outline showing delimbing and topping of standing trees.

Processor

Type of delimbing tool: Trunk embracing knives, two fixed and six moving (Figure 64).

Pulpwood harvester

Type of delimbing tool: Trunk embracing knives, one fixed and two moving. Processing is done when tree is largely in vertical position (Axelsson, 1972).

Infeed: From the side by means of the felling boom, the wholly withdrawn position of which coincides with the direction of the processing unit (Figure 65).

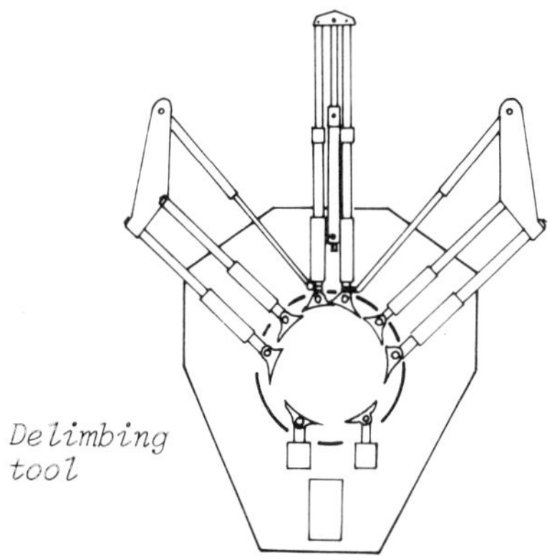

FIGURE 65. Delimbing tool on the processor.

FIGURE 66. Principle outline showing delimbing and bucking of tree in vertical position.

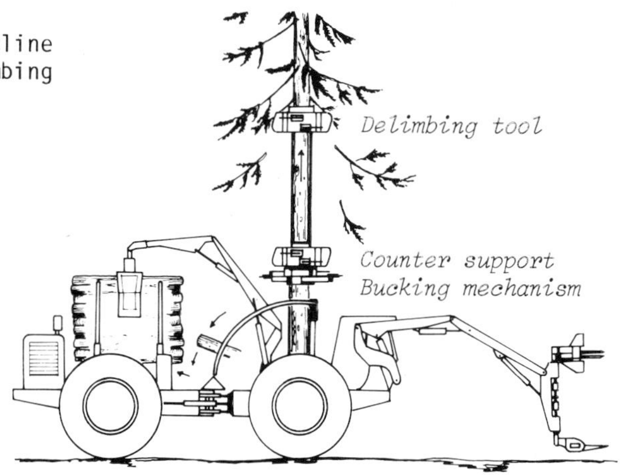

FIGURE 67. Trunk embracing knives and roller feed.

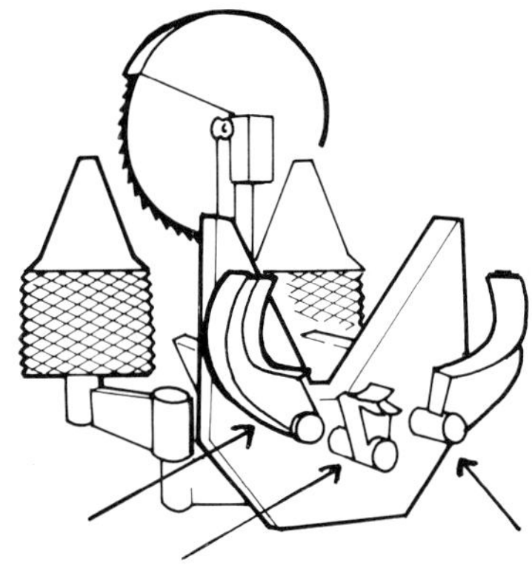

Trunk embracing knives and roller feed

Type of delimbing tools: Trunk embracing knives, two moving tools.

The Garp Rake

The delimbing tools consist of four bow shaped knives with slanted edges. All knives are moving radially. The knives are placed axially so that principally only one limb at a time is removed, even in whorls. Design is shown in a test bench (Figure 69)

FIGURE 68. The Garp rake.

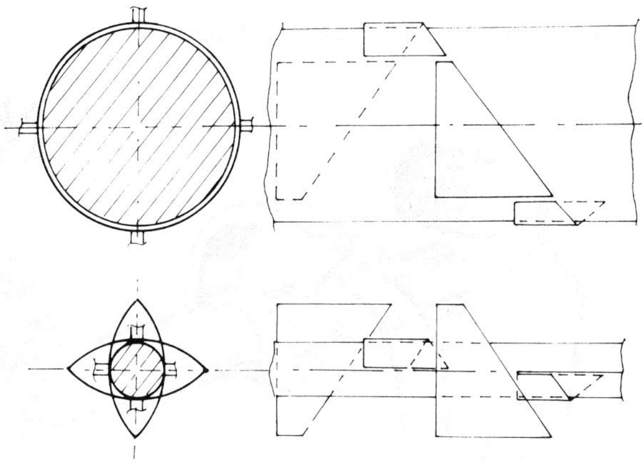

FIGURE 69. The Garp Rake with axially spaced knives which principally cut off only one limb in the whorl at a time. Principle outline of the delimbing tool in a test bench (Staaf, 1972).

Trunk embracing knife track and stepwise feed

Type of delimbing tool: Trunk embracing knife track (Figure 70).

Infeed: Trees felled by means of the machine are lifted into the delimbing tool by means of a boom

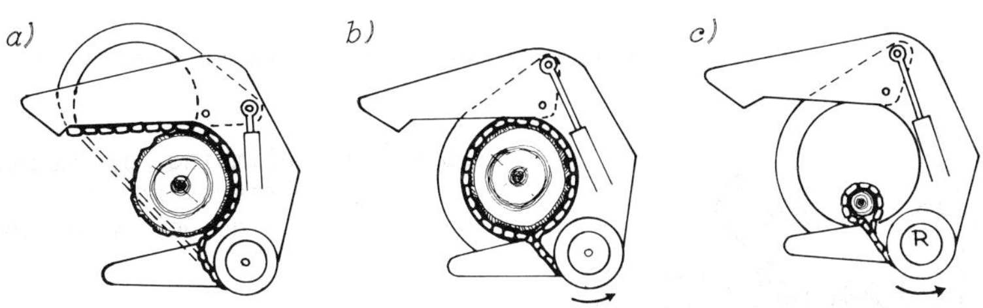

FIGURE 70. The trunk embracing knife track in three different positions. a) tree placed in the knife track. b) knife track embracing the tree. c) at a rate corresponding to declining trunk diameter, the track is wound up on R.

Trunk embracing knife track and roller feed

Delimber

Type of delimbing tool: Knife track (20 knives) with assembly for roller feed and bench including hydraulic cylinders for opening and closure (Figure 71).

Infeed: Grapple gliding on a straight boom

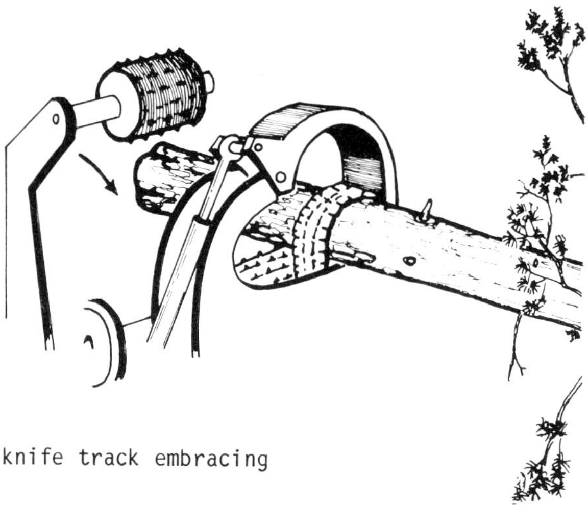

FIGURE 71. Delimber with a knife track embracing the trunk.

Fixed cutters (or corresponding) and roller feed

Delimbing unit

Type of delimbing tool: 8 bolt cutters electrically operated, embracing the trunk (Figure 72).

Infeed: Axial infeed by means of a tractor with boom.

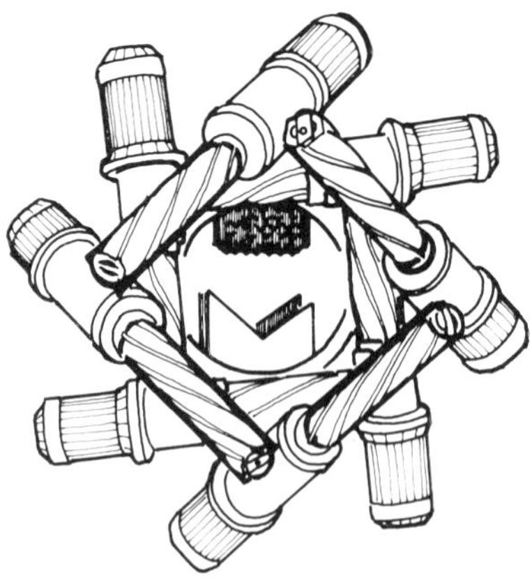

FIGURE 72. Delimbing by means of cutters and roller feed in the delimbing unit.

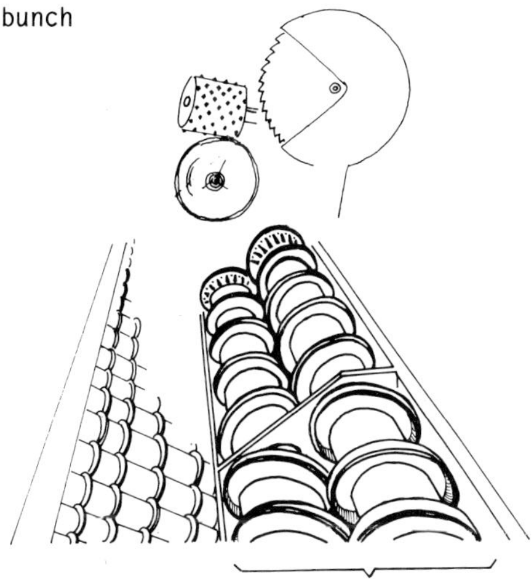

FIGURE 73. Rotating screws in a bunch delimber.

Tools with screws

Bunch delimber

Delimbing tools

The machine consists of four rotating screws with edges along the threads. The screws are imbedded in bearings at both ends, two screws turning clock-wise and two screws counter-clockwise. Logs are placed on an infeed table with screw conveyers for transfer to the delimbing unit. Several logs can be delimbed simultaneously (Figure 73).

Most common delimbing tools

Most common tools used at mechanized delimbing are the trunk embracing knives, often mounted with axially recessed knives for an even distribution of the resistance to delimbing (cf. principle of the Garp Rake, p. 109). The use of debarking tracks has declined largely, these tools being more complicated and expensive in operation than delimbing knives.

Cutters and flails (chains) have disappeared almost entirely in the last ten years. Infeed of the trees at mechanized delimbing may be arranged by means of stepwise feed mechanisms - a short feed between delimbing actions, or continuous feed by means of rollers. In the latter case spike rollers or rubber covered rollers are used for infeed. Because of damage to the wood, spike rollers have been increasingly replaced with rubber covered rollers.

Motor-manual (semi-mechanized) delimbing by means of chain saws is now quite common. In the 70's and 80's the conventional chain saw has been developed into a light and convenient tool for delimbing.

There are now on the market chain-saws with short blades (27 cm) and weighing 4-5 kg. They have an output of up to 2.1 kW (2.9 HP). At harvesting of trees they can be used not only for felling and bucking but also for delimbing, which is then carried out with particular care for valuable saw-timber or special assortments (Axelsson, 1967).

FIGURE 74. A chain saw with automatic brake which not only stops the chain but also eliminates the risk of throws. Max. throw to the hand. Weight including 11" blade: 5.2 kg, Effect: 2.1 kW. This is the only chain saw with a moveable blade.

To reduce time required for delimbing, the types with roller feed will increase in use unless the teeth are judged to be causing considerable damage to the timber. Rubber coated wheels have recently been introduced.

Pulling force of knives and knife tracks varies between 2 and 9 tonnes, depending on rate of feed and tree size. The most common values range between 3 and 5 tonnes. For cutters and similar delimbing tools force is less, or from 1.5 to 6.0 tonnes. However, the lower input required is counterbalanced by greater input required for the efficiency of the cutter functioning (1 tonne = 10 000 N).

The accumulation of limbs in front of the machines may often become a difficult problem, particularly with respect to fixed delimbing tools and at heavy concentration of timber. Arrangements for the removal of limbs are usually lacking. When mechanized delimbing is done by means of mobile machines, difficulties are encountered at the bunching and sorting of timber on the outfeed side.

Conduct of delimbing

Manual delimbing. How delimbing should be done depends on the quality required by the forest industry, i.a. with respect to subsequent debarking. Distinction is made between careful delimbing and simplified (rough) delimbing.

At simplified delimbing 6 cm (occassionally 10 cm) long stubs of the limbs may be left on the trunk.

When tree length trunks are processed, partial delimbing is often applied to the upper side of the trunk wherever possible.

At delimbing of a whole trunk, turning the trunk is frequently a time consuming job. These forms of delimbing apply to manual work.

Mechanized delimbing. At mechanized operations, degree of delimbing achieved is usually acceptable to the forest industries. When delimbing is well done, the industrial raw material is more easily handled and less bulky.

Various places of delimbing. Several factors influence the choice of place for delimbing, such as method of harvest, tree sizes, limbiness of timber, possibilities to combine with other processing, technical means etc.

A systematic arrangement of the options may identify the following places of delimbing:

1. **At the stump**

Standing trees are delimbed:

- manually by means of a pruning knife
- mechanically by means of a tree climbing machine ("tree monkey") or by a multiprocess machine

Trees separated from the stumps can be delimbed in horizontal positions at the stump:

- manually by means of chain saw or axe
- mechanically in horizontal or vertical position by means of a multi process machine.

2. **At strip roads**

Delimbing of felled trees:

- manually by means of chain saw or axe
- mechanically by means of delimbing machines or processing machines for delimbing and bucking

3. **At landings**

- manually by means of chain saw or axe,
- mechanically by means of mobile units or processing machines

4. **At depots**

Various types of stationary units for delimbing can be used more or less centralized in a major production context e.g. at the Russian timber depots to which the trees are transported over long distances i.a. via railways.

Some views on mechanized delimbing

Sensitivity to variations in tree size. When single trees are processed, time required for delimbing is strongly dependent on the sizes of the trees i.e. time required per unit of volume to process small, single trees is considerably longer than time required for large trees. When several trees are treated simultaneously production is almost independent of the sizes of the trees.

Similar to felling, where it is theoretically possible to visualize some form of felling several trees simultaneously in a swath, several trees can be delimbed at a time e.g. in bunches.

- At delimbing of single trees along the trunks i.e. the whorls are processed in succession, a good output is achieved when big trees are delimbed (Tomanić, 1974).
- At delimbing of several trees laterally i.e. the whorls are processed simultaneously along the whole trunk, a good output is achieved when several small trees are delimbed at the same time.
- Delimbing of several trees simultaneously can also be done longitudinally. Such a method has been developed. Small skid loads of trees from thinning can then be pulled through the delimber by means of a skidder winch.

To achieve a high output when big trees or several small trees are delimbed simultaneously, it is necessary to have a high infeed capacity (kW) requiring a sufficiently strong power unit.

Space requirements. Mechanized delimbing of big trees or several small trees simultaneously requires ample space, the operation being carried out by relatively big and heavy machines which need a wide space for the handling of timber.

At mechanized delimbing in clearcut areas and on landings the space requirements of the big delimbing machines and units can be met.

At delimbing in combination with thinning in young stands or on strip roads, space needed for delimbing machines and the choice of methods suited to a high output of delimbing are very limited.

Modern delimbing machines have mostly been developed for operations in clearcut areas. The machines, which are big and heavy, have a high handling capability and strong infeed power units.

Relationships between feeding rate, feeding capacity and infeed power

The rough average relationships based on data from 17 delimbing machines with respect to feeding rates, feeding capacity and infeed power are shown in Figure 75 (Staaf, 1972).

Increased feeding rate requires increased feeding capacity. Infeed power required decreases at increased feeding rate, output being the product of infeed power and feeding rate.

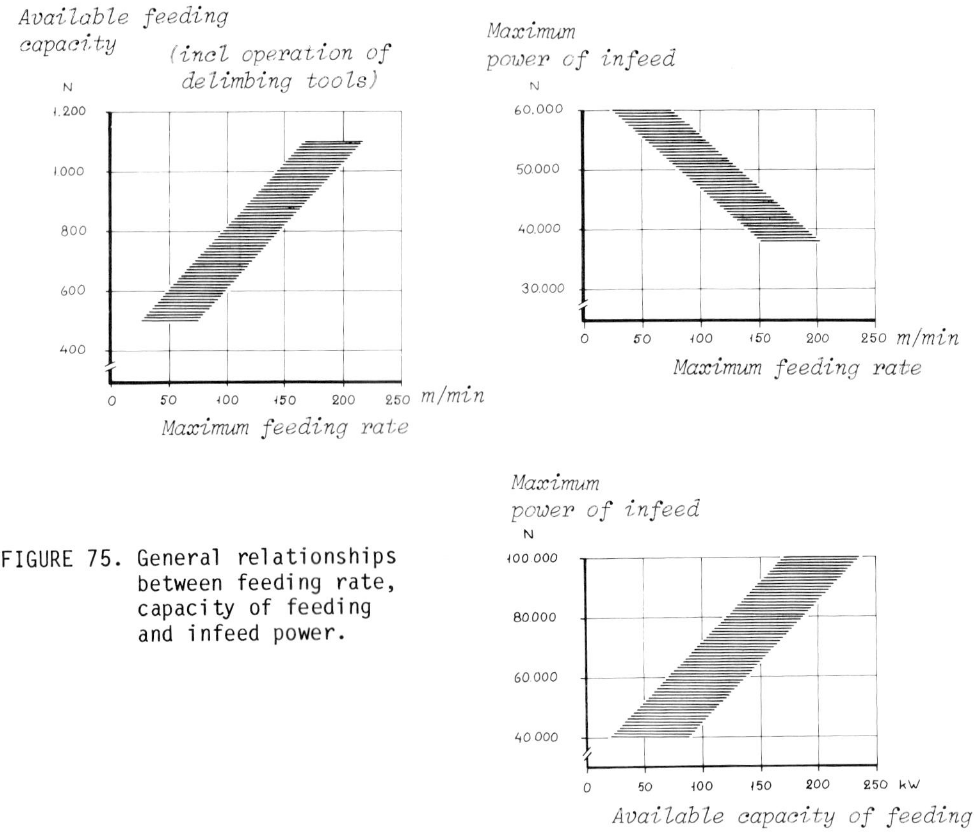

FIGURE 75. General relationships between feeding rate, capacity of feeding and infeed power.

Available capacity of feeding. The fact that increased power of infeed requires increased available capacity of feeding is self-explanatory. However, what minimum available capacity of feeding and lowest maximum power of infeed are required for delimbing of small trees from a thinning operation? This question is of great interest since the size and weight of a multi-process machine for felling and processing in a thinning operation must be severely restricted.

A given capacity of feeding can produce either a high rate of feeding with a relatively low power input or a low rate of feeding with relatively high power input.

Highest possible rate of feeding is desirable at delimbing of small trees in order to achieve a high volume production per unit of time. Infeed power in relation to diameter of limbs is less for small trees than for big trees, limb diameter increasing with age of tree.

Longitudinal delimbing can principally be done in two ways, either by simultaneous delimbing of a whole whorl, or by cutting one limb at a time. In the first case the infeed power required is larger than in the latter case.

Cutting of single limbs. Cutting off one limb at a time can principally be carried out according to Figure 76 which shows the unfolded mantle surface of a trunk with two whorls. The trunk is embraced by four curved knives with slanted edge lines. The edges are placed so that they cut successively the limbs of the whorl, normally 3 - 5 limbs (Staaf, 1972).

Measurements of traction force required. Measurements of traction force required have been carried out in order to elucidate the power requirements at delimbing by means of edged tools. Two types of delimbing tools have been tested, one type having the edge line at an angle of 90° in relation to the direction of delimbing, the other type having an edge line with an angle of 45° for an investigation of the efficiency of a slanted edge. Reduction in force required when a slanted edge was used amounted to 20-30 percent of the force required at a 90° angle (Figure 61).

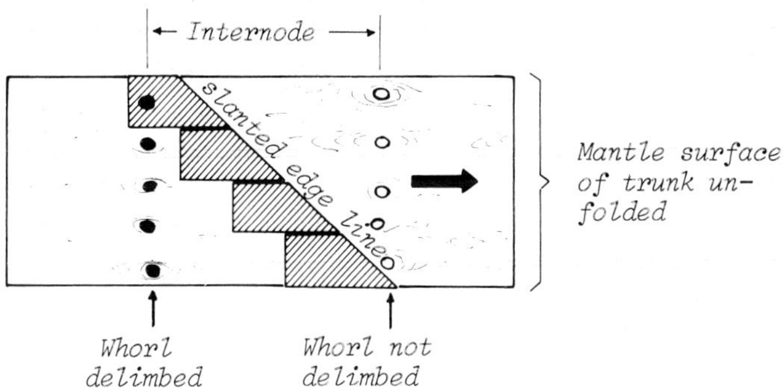

FIGURE 76. The principle design of a delimber cutting one limb at a time.

If the edged tools are put into vibration with a frequency of 60 Hz (fluctuations per second) and a vibration amplitude of 1.5 mm, the force required can be reduced further by 10-20 percent due to reduced friction.

The power requirements can be lowered if delimbing of small trees is done at a varying rate of feed, the rate of feeding increasing automatically for declining diameter, e.g. from a rate of 75 m per min for the 20-cm tree to 125 m per min for the 10-cm tree.

Trees with a diameter of 10 cm and 20 cm may weigh approximately 50 kg and 250 kg, respectively. To bunch these trees in a skidding position, a friction of 200 N must be overcome for the 10-cm tree and 1000 N for the 20-cm tree if half the weight of the tree is assumed to be loaded on the machine and if the coefficient of resistance is 0.80.

If the biggest limb diameter of the 10-cm tree is assumed to be 30 mm and that of the 20-cm tree is assumed to be 40 mm, Figure 61 shows the shearing force required when tools with an angle of 45° are used - see also Table 5 for the theoretical power requirements.

Table 5. Approximate capacity of feeding required at cutting of single limbs by means of an edge at 45° angle (Staaf, 1972).

	Pine		Spruce	
Diameter at breast height, over bark, cm	10	20	10	20
Rate of feeding, m per min.	125	75	125	75
Skidding resistance, N	200	1 000	200	1 000
Shearing force at 90° angle, N	2 500	6 000	4 700	10 000
Total N	2 700	7 000	4 900	11 000
Capacity required, kW	5.3	8.6	9.6	13.4
Capacity required at 45° angle, kw (30 percent reduction)	3.7	6.0	6.7	9.4

Remarks: 1 kp = 10 N (Newton)
1 hp = 0.735 kW (kilowatt)

No allowance has been made for additional force from acceleration of mass or for the mechanical efficiency.

Principally the following approximate relationships are obtained between capacity of feeding and rate of feeding for various sizes of limbs (Figure 77).

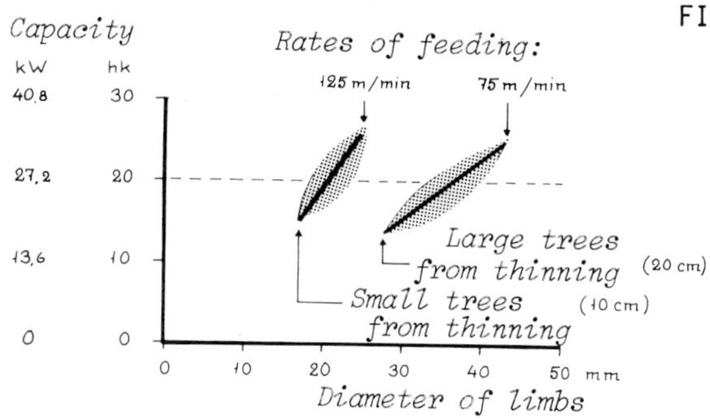

FIGURE 77. Principle relationship between rate of feeding and capacity of feeding at various sizes of limbs (Staaf, 1972).

After calculations of the theoretical power requirements on the basis of data on shearing, an available maximum capacity of feeding of approximately 15 kW (hp) appears to be needed for cutting one limb at a time if the rate of delimbing is varied with sizes of trees and limbs. Actually, this result corresponds to the averages obtained from observations in a test bench.

The average available capacity of feeding in today's delimbing machines used in final harvest operations is 66 kW (90 hp) and the rate of feeding is slightly over 100 m per min.

Quality of delimbing. Requirements for high quality of delimbing must also be met in addition to high rates of feeding at lowest possible power requirements. The delimbing machine must not destroy useful wood in the trunk nor should it leave stubs of limbs that will encumber the subsequent handling and processing.

Quality of delimbing and the design of the delimbing tools. Some principle points of view will be presented here concerning the quality of delimbing in relation to the design of the delimbing tools.

A good allround fit to the trunk can be obtained at maximum tool size if the curved edges are more numerous and independent of each other. When diameter declines towards the top of the tree, the adjustment of the overlapping edges is less efficient and short stubs will be left on the trunk.The stubs, usually of small diameter, do not normally affect the performance and quality of delimbing in a negative way. Such a type of edged tools are used in the test bench for Garp rake according to Figure 78.

FIGURE 78. Four curved edge tools embracing a large and a small trunk.

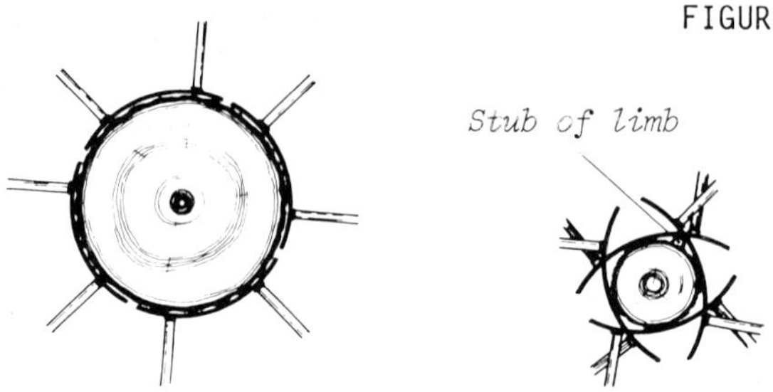

When several curved edge tools are connected to each other it is difficult to achieve a good fit to the trunk at varying diameter. This disadvantage can be alleviated to some degree by delimbing in both directions, as is done by means of machines where the delimbing "wings" and the back of the delimbing carrier are sharpened in both the upper and the lower parts (Figure 79).

FIGURE 79. Delimbing carriage, the delimbing tools of which embrace a small trunk at relatively wide dead angles.

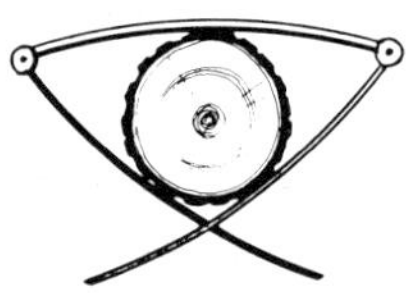

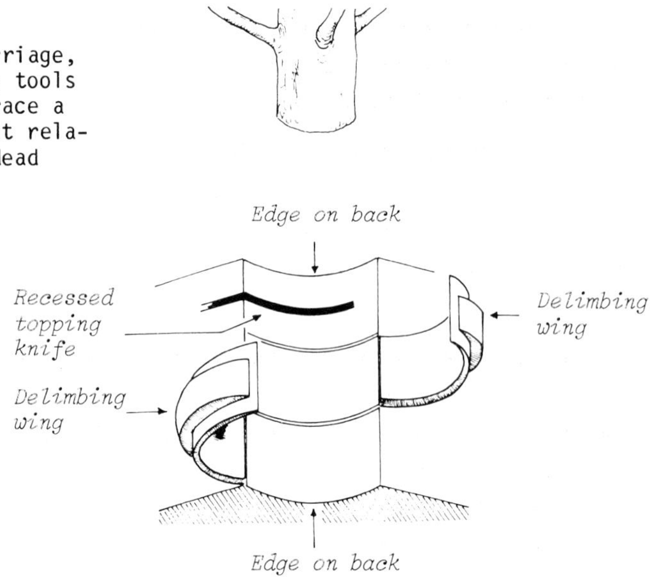

When several straight-edged tools are connected to each other as in a knife track, a good fit is achieved for trunks of varying diameter, with somewhat better result for the big trees than for the small trees. The problems involved in the variation of the diameter are solved by a smooth fit around the trunk and an even pressure by means of the knife track wound up in a fixed point. (See Figure 70).

A good fit around the trunk can be achieved by means of fixed, long and cylindric cutters if several tools are placed at a certain angle to each other. If four cutters are used, the corners are too wide. This can be alleviated by means of an additional four cutters positioned at 45° angle in relation to the first four cutters, creating eight smaller corners or "dead" angles. However, this arrangement gives relatively long studs for the big trunks with big limbs in comparison with the short studs on small trees with small limbs (See Figure 72).

Manual and motor-manual methods of delimbing

The method of delimbing applied depends on the means available, quality of delimbing required, integration of delimbing with the whole processing system and the method of harvesting used.

Delimbing may be manual, motor-manual or mechanized. Quality of delimbing desired varies from careful delimbing to simplified delimbing, or a combination of these forms in partial delimbing.

The methods of delimbing are often different for the tree system, the tree length trunk system and the assortment (short wood) system.

Manual methods. The most common manual method of delimbing is carried out by means of axe, for dry and small limbs also by means of debarking spuds. The use of this method has decreased rapidly in favour of the increasingly dominating delimbing by means of chain saws or more modern equipment.

Motor-manual methods

Delimbing of felled trees. Motor-manual delimbing is predominantly done by means of chain saws. It is important that a well thought-out and practiced technique is used. Work is largely facilitated if the tree trunk is used as support for the chain saw.

A systematic delimbing of whorl after whorl, alternately cut from the right to the left reduces to a minimum the path that the chain saw must travel through the crown. In this work, the delimbing operation can be divided into six different steps that are carried out from one position.

Figure 80 shows the techniques for delimbing of the upper side or partial delimbing for coarse limbed trees e.g. the leverage technique.

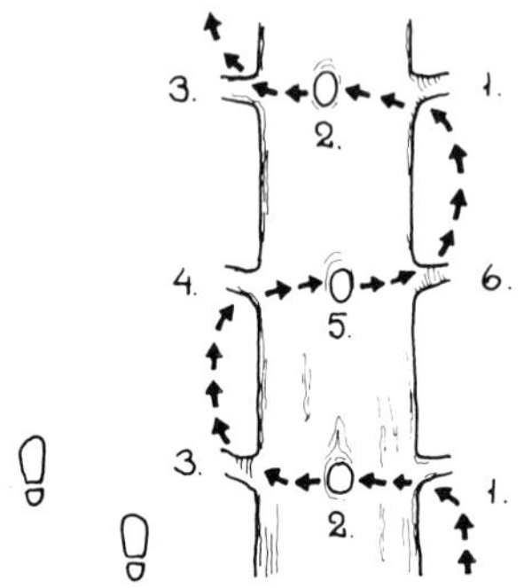

FIGURE 80. The principle of delimbing the upper side according to the leverage technique.

The leverage technique

Upper delimbing. Upper delimbing by means of the leverage technique starts at limb no 1 using pushing chain action and keeping the chain saw against the trunk and the right leg against the left side of the trunk for good balance. The chain saw is then turned over the trunk so that the bar rests on top for cutting limb no. 2. The chain saw still operates with pushing chain action.

The chain saw is positioned for cutting limb no. 3 with pulling chain action by using the right knee as support. For limb no. 4 the chain saw is moved forward still supported by the right knee for cutting with pushing chain action.

The chain saw is then placed on top of the trunk for the cutting of limb no. 5 by means of pushing chain action. At last, limb no. 6 is cut by means of pulling chain action after the chain saw body has been turned up for support against the trunk. The feet are now moved to a new position for the next pair of whorls.

Delimbing of lower side. After the upper delimbing is finished, delimbing of the lower side is done, usually quickly and safely.

At a rather normal level of the trunk at knee height (50-70 cm) and after the still limby side of the trunk has been turned 90°, delimbing is done using the trunk as protection and support for the chain saw from the top of the tree towards the butt end. The saw is operated with pushing chain action.

When the trunk is resting on the ground, the limby side is turned up to become accessible for delimbing by means of pulling chain action in order to reduce the risks of throws and sawing into boulders.

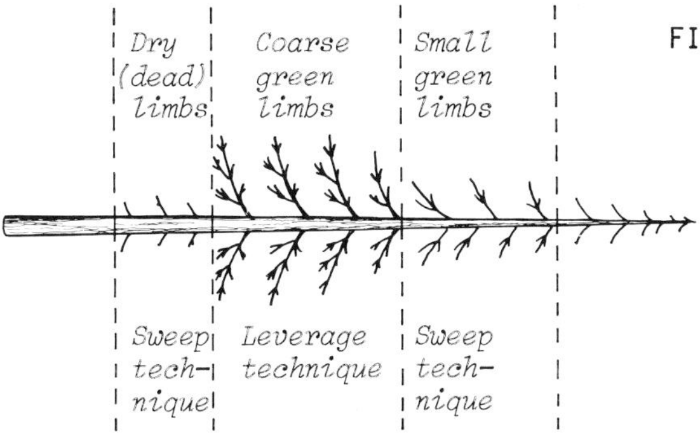

FIGURE 81. Principle of delimbing by means of the sweep technique.

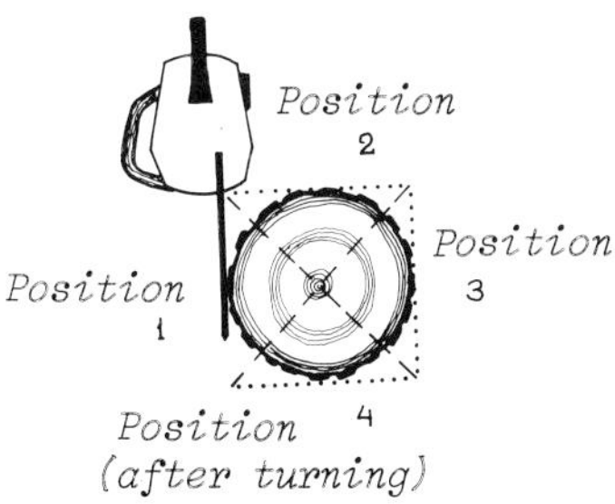

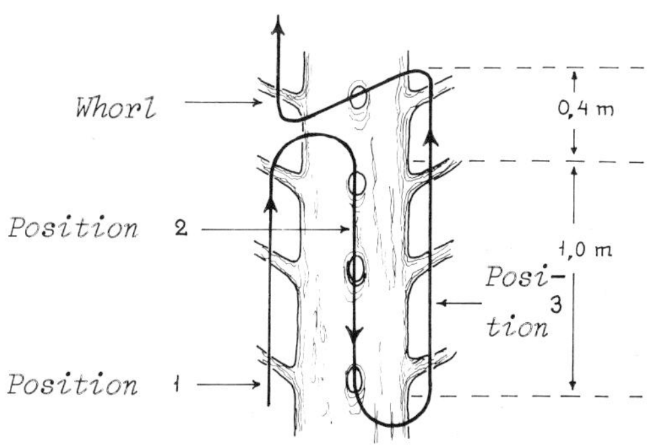

Sweep technique. The sweep technique is used for small limbs where the chain can cut rapidly each limb or several limbs simultaneously or in succession during one single sweep. The path of the chain saw through the crown is longer than with the leverage technique since the distance between the limbs in the whorls is usually shorter than the internode between the whorls.

The gain in using the sweep technique is obtained from the fast progress through the crown.

Figure 81 shows for the sweep technique how the chain saw is placed against the trunk and moved forward and backward in metre-long sweeps.

When the sweep technique is used, it is preferable to operate with pushing chain action at low height above the ground and with pulling chain action at higher level above the ground in order to obtain a more convenient working position. On the upper side of the trunk both pushing and pulling chain actions are used.

To retain a convenient working position, delimbing of the whole trunk should be carried out before bucking if the trunk can be turned. This is recommended procedure in particular for big trees which also require more careful scaling. When the trees are cut into pulpwood of standard length, scaling and bucking can be done simultaneously with delimbing more quickly.

FIGURE 82. Delimbing of a tree felled across a base tree. A chain saw is used for delimbing.

Sweep technique and leverage technique combined. A combination of the sweep technique and the leverage technique can be advantageous at delimbing of long tree crowns with varying sizes of limbs.

Some safety rules at delimbing

- Use personalized protective equipment
- Acquire safety equipment for the chain saw
- Keep stable and safe position of the feet
- Never change position of feet if saw is held on the left side of the trunk
- Always keep right leg behind the front handle, the thumb underneath the bar
- Keep close to the chain saw

- Avoid cutting by means of the saw tip, use safety chain and protection against throws
- Use a well sharpened chain
- Use a chain bar suitable for the size of trees that are to be processed
- Keep close attention to the movements of the tree and the limbs
- Check that the working position is safe.

Delimbing of standing trees

Climbing machines. Experiments with climbing delimbing machines for standing trees carried out in recent years have produced a number of new designs.

In view of current trends in the development of labour reducing methods and considering the requirements concerning the weight, performance and costs of climbing machines (Denmark), it has been judged that the prospects of a wider use of this motor-manual method of delimbing are rather slim. A climbing machine would hardly find an acceptable function as an integrated part of a harvesting unit e.g. a feller-buncher. Its delimbing performance is all too low in comparison with that of telescopic delimbing machines or similar devices.

Still in the 1980's climbing delimbing tools have been developed primarily for the improvement of the quality of coniferous trees. Two types can be distinguished viz. a hydrostatically operated unit and an engine powered unit.

The power unit of the hydrostatically operated type is usually a tractor mounted hydraulic pump to which the delimbing unit is connected by two hoses. The unit, which also has two trunk embracing knives, is operated with a gentle pressure against the bark by means of rubber covered wheels. Rate of progress may exeed 5 m/s. Delimbing upwards can be done to a height of 25 m. Supplementary delimbing may be carried out on descent. Shearing power on limbs amounts to 5000 kp (5 tonnes). A machine of this type weighs approximately 40 kg for an engine output of 12 kW. It is manufactured in West Germany.

Another type has been developed and manufactured in Japan. It is a climbing unit with a rotating engine powered chain saw. It is operated with four wheels pulling and six caster idler wheels. Output at work is 2.25 m/min. The unit weighs 25 kg. It is possible for one person to operate two machines at a time. The unit can be operated from the ground by means of a rope (cable) connected to the engine choke (Toyorinno, 1983).

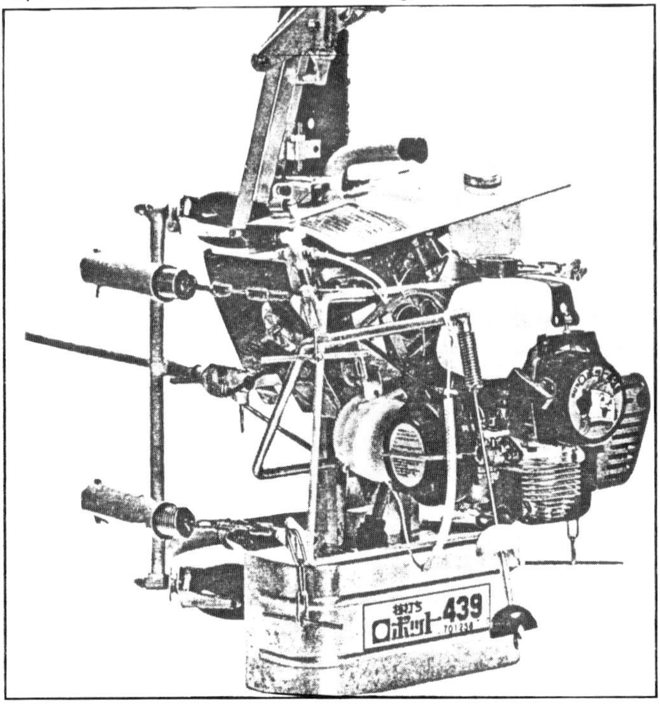

FIGURE 83. Tree pruning machine.

Mechanized delimbing

A large number of different delimbing machines or processing machines with integrated delimbing operation have been developed for mechanized delimbing in order to eliminate the labour intensive manual delimbing work.

On the basis of identified places of delimbing some typical machines will be presented here for delimbing at the stump, at strip roads and at landings.

Delimbing of felled trees in horizontal position at the stump

One of the first multi-processing machines for harvesting of trees was ready for practical use in 1960. In addition to felling and delimbing it could also carry out bucking, bunching and transport of the timber to the truck or to a trailer. It was first tested in USA at harvesting of yellow pine.

This machine can harvest approximately one tree in two minutes (average tree size: 24 cm at breast height), 25 m^3 (piled) per day. Under favourable conditions it may process up to 7.2 m^3 (piled) per hour of efficient work.

Delimbing of trees in vertical position after separation at the stump

A Canadian machine which is built on a tractor with hydraulic 4-wheel drive, carries out felling, delimbing, topping, bucking into 2.5 m bolts, and transports 17 m^3 of pulpwood loads to landings.

Performance of machine. The output of the machine is approximately 6.5 m^3 per hour of efficient work. The machine is operated in two 8-hour shifts, 5 days per week.

Delimbing of trees in vertical position before felling. A tree harvester is chosen to represent this method of delimbing.

The machine can handle delimbing, topping, separation from the stump, and piling of trunks. It operates in the stands.

Design of the chassi. The chassi is principally built as that of an excavator. It consists of a frame and a revolving upper part equipped with knuckle boom and a mast carrying mechanism for delimbing, topping and separation from the stump. The mast consists of a fixed (inner) part and a moving (outer) part, felling shears and delimbing carriage.

The felling shears are attached to the lowermost section in the fixed part of the mast and the delimbing carriage glides on the moving part of the mast.

The delimbing carriage consists of a frame, delimbing braces, topping knife and hydraulic components. Two sizes of masts and processing mechanisms are manufactured.

As is the case with other harvesting machines, this tree harvester has its greatest potential in mature stands with big trees.

Since the machine is capable of delimbing up to a height of 18 m only, processing in stands with big and tall trees constitutes a problem. However, this is technically a small matter of size.

Delimbing at strip roads. A delimber - buncher is chosen here to represent delimbing at strip roads.

Method of work

The machine is designed for final harvest stands. Proceeding backwards along a strip road, it delimbs all trees that can be reached from each position, the telescopic boom having a maximum reach of 12 m from the pivot.- For a minimum tree height of 10 m the machine is expected to operate from strip roads laid out 30-40 m apart (Figure 84).

To accommodate the machine, felling should be arranged so that the largest possible number of trees can be reached by means of the telescopic boom from each position. This can be achieved if:

- the trees can be grappled in the crown at the top end of the merchantable timber when delimbing is done from the top to the butt end.
- the trees and the telescopic boom are aligned as closely as possible in order to facilitate the attachement of the delimbing grapple on the boom to the tree.
- the butt ends of the trees are not blocked causing time consuming pulling apart
- the predominant direction of wind is considered
- delimbing can be done in positions along the strip road which accommodate the positioning of the machine and terrain travel.

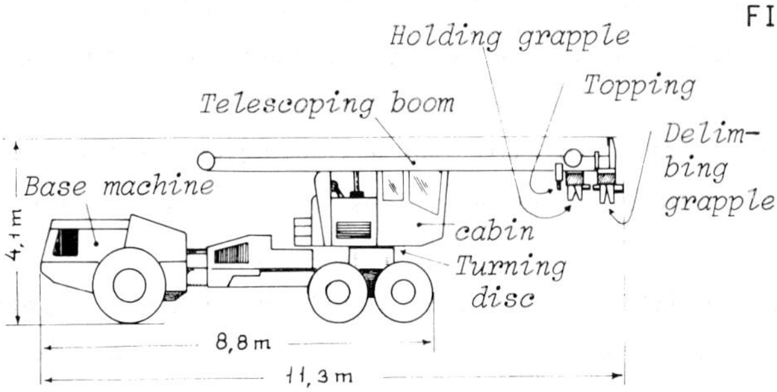

FIGURE 84. Machine for delimbing and bunching.

Process of delimbing

The process of delimbing can be described in the following way. The boom is extended and the delimbing grapple embraces the tree at the top end of the merchantable timber. The boom is then pulled in to the holding grapple which grabs the tree at the end of the merchantable timber. During this movement about 1 m of the trunk at the end of the merchantable timber is delimbed by means of the fixed mechanism on the holding grapple.

Behind the holding grapple there is a hydraulic grapple for bucking the top when the tree is pulled in. Delimbing is then done by means of the edged tols on the delimbing grapple which is extended from the machine along the tree.

If the maximum telescopic movement (7 m) is not sufficient for a complete delimbing, a retake is made by means of the holding grapple. When

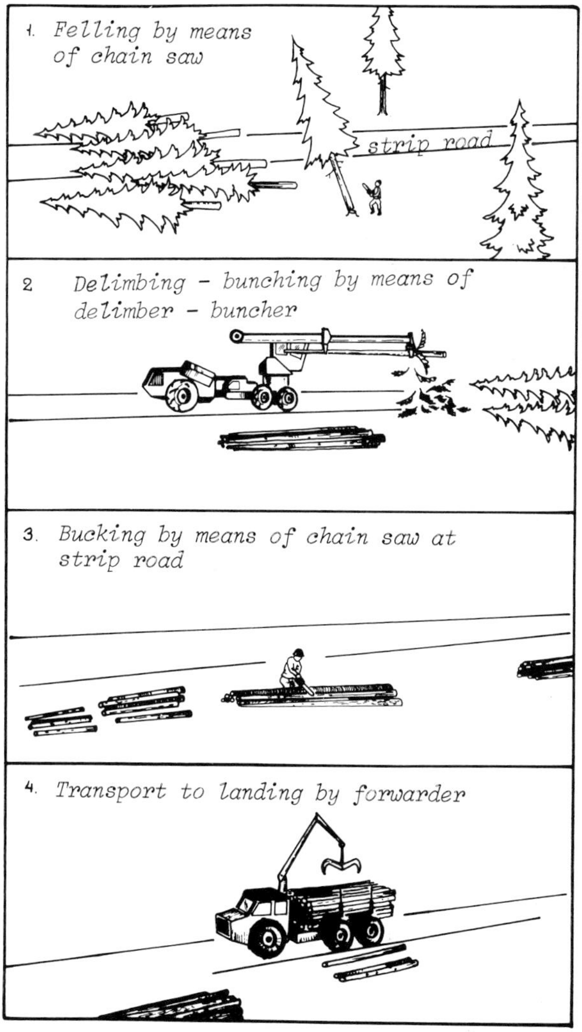

FIGURE 85. Delimber-buncher for processing of assortments (short wood) at strip road.

delimbing of the butt end is complete, the holding grapple grabs the trunk, the delimbing grapple is released and the boom is extended.

After delimbing, the trunks are placed along the strip road in order to facilitate rational handling during the following processing e.g. bucking into assortments and piling or even-end bunching for transport to a place of bucking (Figures 85 and 86). The trunks may be sorted by species.

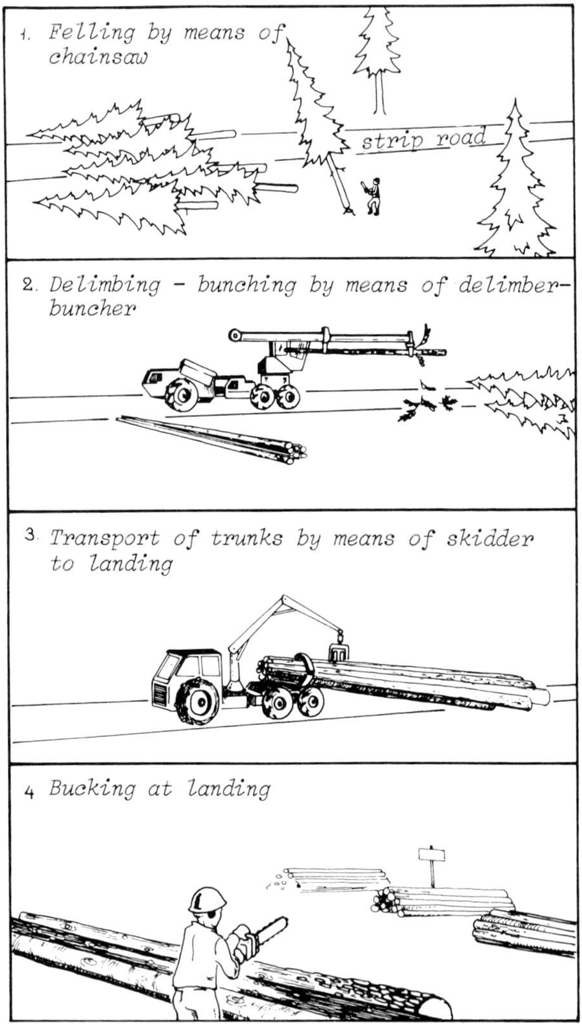

FIGURE 86. Tree length trunk method

Output of machine

Depending on the stand conditions, the machine output varies between 50 trees and 130 trees per hour of efficient work for big timber (average 0.65 m^3) and small timber (average 0.10 m^3), respectively.

Delimbing integrated with other harvesting operations

Mechanized delimbing, bucking and bunching at strip roads

Several machine types have been designed and developed for the combination of mechanized delimbing, bucking and bunching at strip roads.

Work procedure of the machine

The processor delimbs, bucks and bunches trees felled in final harvest operations. The machine is programmed and maneuvered by one man. It is built on the chassi of a forwarder with a rear bogie above which the processing unit is mounted on a pivot allowing 270° turns.

The infeed boom, which is turning together with the processing unit, has a gliding (parallel) boom with a reach of 12.6 m. The trees are pulled by means of the boom into an infeed bunk (bench) where the butt end is held firmly until the delimbing process starts. The grapple of the boom is equipped with edges for a rough delimbing carried out when the straight boom is extended along the trunk for infeed. Due to the rough delimbing, the loose limbs are spread out, reducing the amount of slash in front of the delimbing unit.

The tree is brought from the infeed bunk into the processing ramp by means of a boom arrangement (Figure 87).

Delimbing tools

Delimbing is done by means of two knife tracks and through-feed is obtained between two power drives, hour glass shaped spike rollers, underneath the tree, which are pressed against an upper drive pressure roller.

Mechanized delimbing and bucking at landings

Single-tree and multi-tree machines are used for mechanized delimbing at landings. An example of the single-tree machines is a delimbing machine included in a depot and an example of the multi-tree machines is a bunch delimber.

FIGURE 87. Delimbing and bucking by means of a processor

Delimbing depot

The delimbing depot is a mobile establishment of delimbing, bucking and sorting at landings. A complete depot consists of seven units: delimbing machine, infeed machine, limb conveyor, bucking and sorting unit, extension table, maneuvering cabin and electric unit.

Felling and transport to the delimbing depot

Forwarding of trees from the stumps to the delimbing depot is done by means of articulated wheel tractors with clam bunks or winching equipment.

Felling machines of the type feller-skidder are used for transport of trees when the entire harvesting system is mechanized.

Output

Output varies strongly with the sizes of the trees. When 2-3 small trees are delimbed simultaneously, the adverse effect of small sizes is slightly counter-acted. Normally, 90-200 trees per hour of efficient work are delimbed.

Economically, the best result is obtained at delimbing of old trees, particularly spruce, with long and limby crowns. Th depot, therefore, is used primarily at final harvest operations for very coarse trees, difficult to delimb.

The depot is less advantageous for delimbing of trees with short crowns.

Principle of delimbing

Delimbing is done by means of eight cylindric cutters embracing the trunk. Each cutter has its own motor with a momentary output of 7.5 kW. It is guided by a rod which is gliding on the mantle surface of the trunk, determining how close to the surface of the trunk the limbs should be severed by the cutters.

Normally the guiding rod is set to produce a delimbing that cuts the limbs 5-7 mm above the mantle surface of the trunk. The stubs will be slightly longer in the "dead" angles between the cylinders, and longer on big trunks.

Each cutter and guiding rod is held against the trunk at a constant, light holding pressure by means of a hydraulic cylinder. This will keep the cutter following automatically the more or less even surface of the trunk. Valuable wood in the outer part of the trunk would be lost if there were no guiding rods. The rotation and forward slanted placement of the cylindric cutters are pushing the limbs to the side where slash will be taken care of by a conveyor. This design will keep the machine free from limbs.

Bunch delimber

The bunch delimber is primarily designed for delimbing of small timber in bunches. It is equipped with a bucking saw in order to facilitate processing of timber into length desired before the delimbing operation starts. The machine is mobile and designed for one operator.

Work procedure of the bunch delimber

A loader puts a bunch of small trees (2-10 trees) with a volume of about 1 m^3 into the infeed mechanism which is compressing the bunch and breaking the limbs. The whole bunch is then fed into the delimbing unit.

Between the infeed and the delimbing unit there is a circular saw bucking the trees into maximum 5 m length. The operator handles this work from an insulated, well protected cabin by means of push buttons. The moving parts are maneuvered hydraulically.

The delimbing machine treats the bunch as if it were a single trunk. It can also delimb single trees. When the trees are big the machine is giving a particularly good production if used for single tree operation. Note the opposite situation with the delimbing depot. However, the machine is designed primarily for small timber (Figure 88).

The machine bucks and delimbs rapidly and with great capability. The scaling procedure is rough but usually sufficient, the product being mainly pulpwood.

The cycle of operation is approximately equal for bunches and single big trees. The production, therefore, is relatively independent of tree sizes.

FIGURE 88. Bunch delimber.

Output

In practice output varies between 150 m^3 and 250 m^3 of solid wood per shift, keeping two loaders fully occupied. The machine can delimb trees with a diameter at breast height of up to 75 cm and a limb diameter up to 7.5 cm. The limbs are cut into pieces of 5-20 cm length by means of the rapidly rotating knife rollers. Two of the rollers are threaded clock-wise and two rollers counter-clockwise in order to prevent the bunch of trees from shooting off axially.

Trends in delimbing

Some views on the weight of delimbing machines

A mobile harvesting machine, particularly for thinning should not be bulky nor heavy. This requirement means that all components of the machine and, hence, also the delimbing unit should be of light weight.

The weight of trees in thinned stands is relatively low (50-150 kg) while weight of trees in final harvest stands varies between 500 kg and 1000 kg. It should therefore be less difficult from a weight-design point of view to coordinate the partial operations in a continuous parallel production at thinning than it is at final harvest operations.

The possibility to integrate the partial operations has been greatly improved due to the technical evolution in recent years with respect to new machine components and refined control techniques such as pneumatic control of opening and pressure.

The weight of a delimbing machine can be reduced if light edge tools are used instead of heavy cutters and flails with associated power units, and if roller feed can be applied.

Increased mechanization of delimbing can be expected

Quality of delimbing by means of current techniques is usually very dependent on diameter of the trunks. The relative time required for delimbing is high for small trees which are also considerably more limby than old and big trees.

Development of tree methods and tree part methods when limbs and tops are processed for fuel purposes has provided new opportunities for an improvement of delimbing techniques. Lowered requirements in the pulpmills for high quality of delimbing in recent years have also changed the situation regarding the methods of delimbing.

There are a large number of various technical solutions to the problems in the mechanization of delimbing. The solutions can be classified into systems according to various principles. The following figure shows a division of the delimbing principles (Dahlin, 1983).

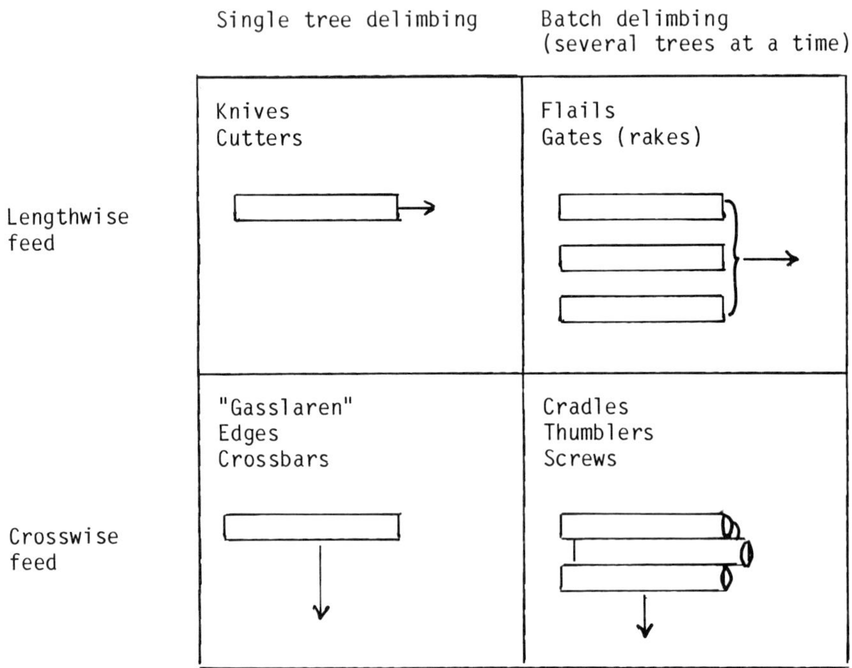

Figure 89. Division of the main principles of delimbing.

1. Output capacity at delimbing of single trees by lengthwise feed is proportional to the trunk diameter squared, and rate of feed.

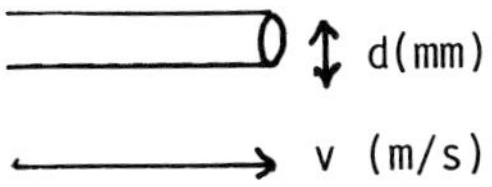

$$P(m^3/h) = f(d^2 \cdot v)$$

FIGURE 90. Output capacity (P) in relation to trunk diameter (d) and the rate of feed (v) at delimbing of single trees by lengthwise feed.

2. Output capacity at delimbing of single trees by crosswise feed also depends on diameter of the trunks if distance between each tree or part of tree is constant. If the trees or the tree parts are packed together, output capacity is proportional to the diameter of the trees or the trunks.

a) $P = f(d^2 \cdot v)$

b) $P = f\left(\frac{d^2 \cdot v}{d}\right) = f(d \cdot v)$

FIGURE 91. Relationships between output capacity (P), diameter (d) and rate of feed (v) at delimbing of single trees by crosswise feed. a) constant distance between the trees (trunks) or b) distance between the trees (trunks) depends on diameter.

3. Three different principles can be distinguished at delimbing of several trees (trunks) by lengthwise feed.

   3.1 A constant number of trees in the batch. Output proportional to diameter squared.

   3.2 The trees are put in layers. The number of trees in each layer depends on space available. Output of delimbing directly proportional to diameter of the trees.

   3.3 A bunch of trees with a given cross-section area is delimbed. Output depends on the rate of feed only.

a)

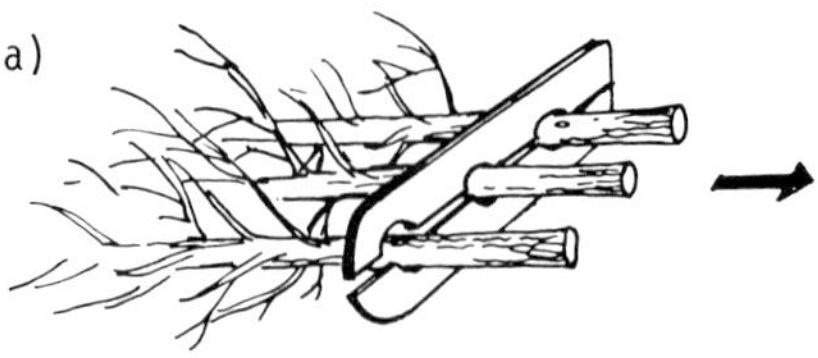

$P = f(d^2 \cdot v)$

b)

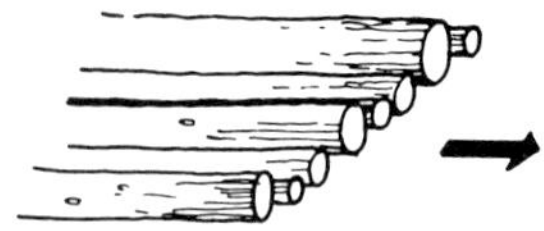

$P = f\frac{d^2 \cdot v}{d} = f(d \cdot v)$

c)

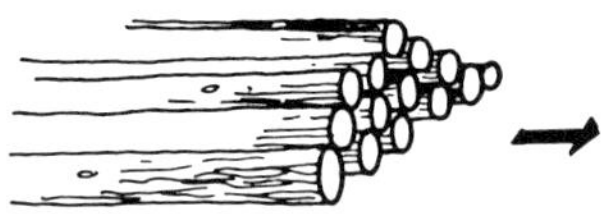

$P = f\ \frac{d^2 \cdot v}{d^2} = f(v)$

FIGURE 92. Relationships between output capacity (P), diameter of trees (d) and rate of feed (v) at delimbing of several trees by lengthwise feed. a) constant number of trees in each batch. b) number of trees in a layer depends on diameter. c) bunches in which the number of trees depends on diameter of the trees.

4. Delimbing of several trees by crosswise feed.

A batch of trees is delimbed for a given time after which a new batch is processed. Output is proportional to the rate of feed, i.e. time given for delimbing of one batch.

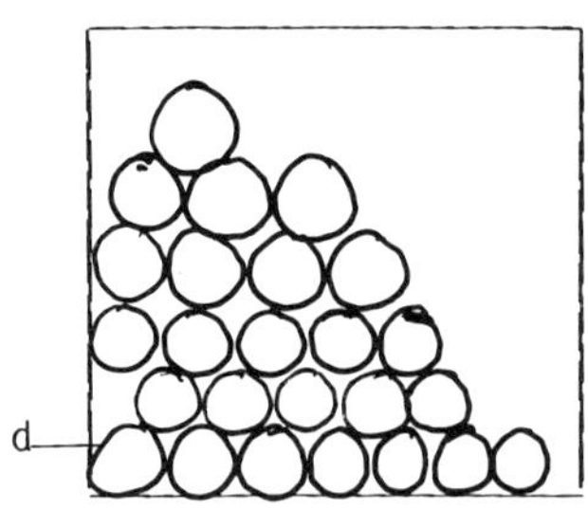

$P = f\left(\frac{d^2 \cdot v}{d^2}\right) = f(v)$

v = time per batch

FIGURE 93. Relationships between output capacity (P), diameter (d) and time or processing per batch (v).

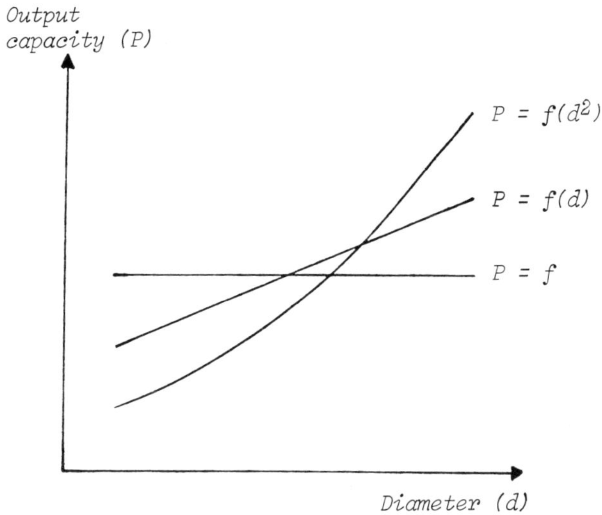

FIGURE 94. Graph showing the influence of diameter on output capacity at constant rate of feed (Dahlin, 1983).

When single trees are delimbed by lengthwise feed, the rate of feed should be varied in relation to the diameter of the trees. Small trees can then be delimbed rapidly and output is increased.

Delimbing several trees at a time (batch) by lengthwise feed can be achieved by means of edged tools through which the trees are pulled, or by a simultaneous delimbing by means of grader blade mechanisms pushed or pulled along the trees. Another method is using the flail techniques by which chains or flails mounted on a rotating axle remove the limbs. More recent methods may cause timber defects.

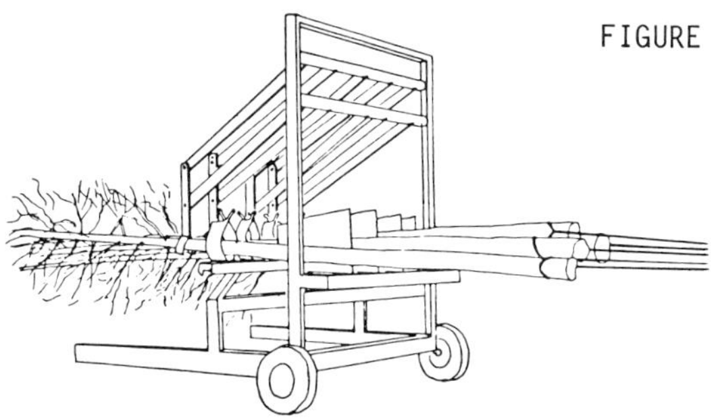

FIGURE 95. Raking delimber. Trunk embracing knives, feeding by means of winch (Taraldrud 1972).

FIGURE 96. Hydro Axe 500, delimber with flailing chains.

Delimbing of single trees by crosswise feed should give a high output. The rough timber is fed between small plates that move at various speed. The timber is brought into rotation when it is moved forward. Delimbing is achieved by means of edges or cross-bars that cut or break the limbs. This method has recently been tested at the Faculty of Forestry in Sweden.

FIGURE 97. Outline of the basic principle applied in the delimbing unit.

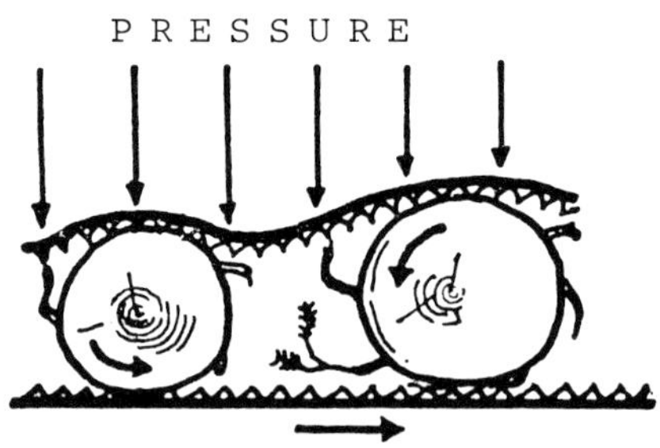

Delimbing of several trees by crosswise feed can be carried out by means of different varieties of delimbing machines equipped with e.g. long rotating rollers with spiral shaped edges (thread), various types of cradles where the batch of trees is rotating or thumblers through which the timber is passing continuously.

In all of these cases friction contributes to the delimbing process. Quality of delimbing depends on duration of processing.

Recovery of residues from delimbing provides an opportunity to improve the economic result of tree harvesting. This can be done most efficiently when the tree or the tree parts are delimbed at a major terminal or at industry. Large delimbing units operate with high output, processing several trees at a time by crosswise infeed. Small delimbing units lend themselves to single tree processing by crosswise feed.

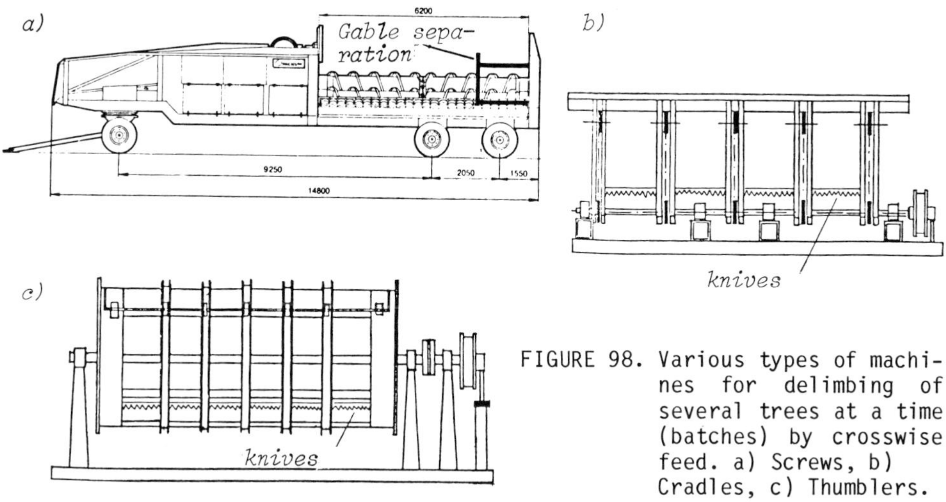

FIGURE 98. Various types of machines for delimbing of several trees at a time (batches) by crosswise feed. a) Screws, b) Cradles, c) Thumblers.

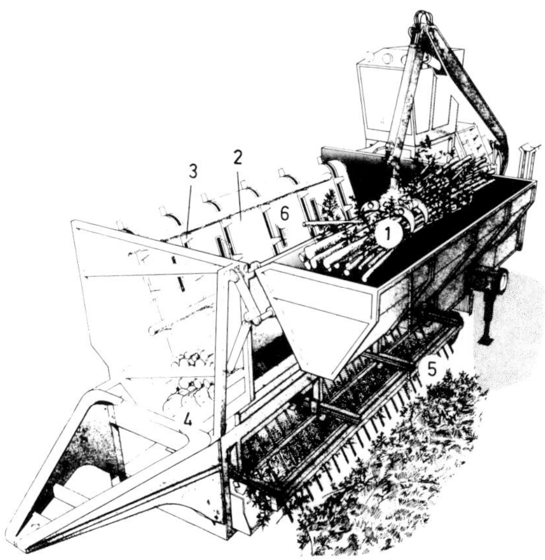

FIGURE 99. Description of bunch delimber in operation.

Short storage of tree parts in the infeed bin (1) until the previous batch has been delimbed and discharged from the through (2). Infeed bin then tips the bunch into the through where the outfeed conveyor tags (3) put the bunch into rotation. The processing rollers break and grind the limbs into a mash that falls down between the rollers and is ejected to the side of the scraping conveyor (5). During the delimbing process the push-off pins (6) are in position to allow the tree parts to fall into the bundle. Delimbed parts are fed out of the through when the push-off pins are retracted. (Skogsarbeten, 1983).

Bucking

The purposes of bucking a trunk are primarily the following:

- to separate various assortments from each other (sawlogs, pulpood, special assortments of various kinds)
- to produce feasible handling units for i.a. transport and other forest operations and at industries
- to obtain certain practical measurements of timber for its scaling and marketing

Similar to delimbing, bucking can be done manually, motor-manually and mechanized more or less automated at the stump, at strip roads, landings or at industry.

Importance of bucking

The bucking operations determine the size of the handling units.

The size of the handling units varies geographically due to variations in the average size of the trees and due to differences in the forms of handling and transport used.

Concerning lenght of timber the situation is changing gradually with the introduction of mechanized harvesting methods and with the restructuring of transports. The trend has been, and still is, an increased bucking of timber into a standard length of 3 m at the expense of random length and 2-m length. This applies to pulpwood.

Tools and means of bucking

Manual tools

Axe. Until the latter half of the 1800's the axe was used for bucking big trees.

The axe is still used for bucking of tree tops when small trees are being delimbed, simply by one or a couple of well aimed chops.

Log saw, 1-man saw, bow saw. Prior to the development of the chain saw, bucking was usually done by means of some type of hand saw. The log saw for two men was used at the end of the 1800's and in the beginning of the 1900's. It was replaced by the 1-man saw for bucking of big trunks and by bow saws for small trunks.

Motor-manual tools

Portable saws with motor powered chains. This category includes the saws we call motor saws, power saws or chain saws. An estimated 75 percent of all timber harvested is now bucked by means of chain saws. Most of the bucking is done near the stump while other bucking is done at strip roads and landings.

Portable motor powered circular saws. Circular saws are primarily used for motor-manual bucking of small timber. This method of bucking is used to a very limited extent.

Mechanized bucking

Chain saws. Chain saws are primarily used at processing depots e.g. at the delimbing depots. The chain saws may be powered mechanically, electrically or hydraulically.

Circular saws. Circular saws of relatively large sizes are used for bucking of timber at strip roads in mobile processing machines and at processing depots on landings.

Shears. Hydraulic shears have been introduced for mechanized bucking primarily in processing machines.

Machine power. Previous physical work in forest operations has been gradually replaced with machine work. A good forest worker may achieve an output of 73 W (0.1 hp), momentarily sligthly more. A horse can sustain a continuous pulling force of 1000 N (100 kp), momentarily 3000 N (300 kp) over a short distance (approx. 60 m).

If a forest worker weighing approximately 70 kg is compared with a motor saw (weight 7 kg) with an output of approx. 3 kW, it is obvious that the lifting and pulling power available by means of machines is much greater than that of physical work. Hydraulic cranes can lift 1 tonne at a distance of 5 m from the pivot (5 tonne-metres). Tractors may have pulling forces exceeding 150 000 N (15 000 kp), which corresponds to that of 150 horses.

Advantage of machine power

It is now possible by machines to achieve outputs of 1000's of horse power. Compare the output that living creatures, man and animals, can produce, i.e. approximately 0.1 kW per 100 kg of body weight, with the output of 0.6-6 kg per kW for a motor, and 60-120 kg per kW for a machine e.g. tractor.

A further comparison between living creatures and machines gives rise to the following comments:

- the machine knows no fatigue
- the machine does not require the same care as a man or an animal
- the machine can operate continuously for long periods of time
- the machine can work with greater precision than labour
- the machine can carry out whole work procedures automatically
- the machine assists Man in raising productivity, releasing Man for other work and giving Man time for own disposal
- the machine can replace physically heavy work
- machine operation, however, requires a higher level of education

The revolutionary changes in the power resources available in forest operations explain e.g. the contemporary change of handling and processing of timber.

The importance of bucking for the production of feasible handling units has been mentioned. The importance of bucking for the achievement of the highest possible value from the tree will be discussed in the following presentation.

Bucking is done after scaling, or marking the length of various assortments. Scaling takes into consideration diameter and length of trunks, weight of loads, quality requirements, species, occurrence of knots, defects etc.

Various methods of bucking

Bucking by means of various equipment can be done in three different ways, using stationary, mobile or moving bucking equipment. The distinction given here is primarily meant for bucking depots or mechanized bucking arrangements.

Stationary bucking equipment

The stationary bucking equipment is mounted in a fixed position in a depot or at the end of a conveyor. The stationary bucking unit can be a circular saw, chain saw or hydraulic shears. The circular saw is usually mounted in a swing, the working cycle of which consists of the partial movements: approach, bucking and return.

An approach speed of 0.5 m/sec and a rate of bucking of 0.3 m/sec by means of a circular saw is shown in the graph (Figure 100).

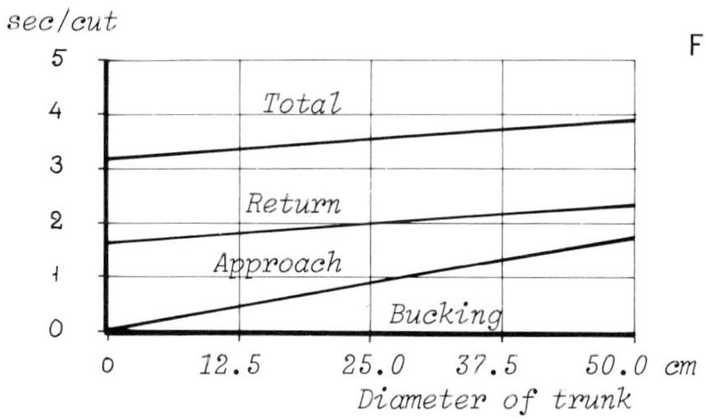

FIGURE 100. Graph showing bucking time for a circular saw cutting trunks of various sizes. Fixed unit. (Arnelo and Banner, 1967).

Mobile bucking equipment

The mobile equipment that bucks when standing still can be a chain saw. It is mobile along the conveyor within certain limits and it can be used at depots for processing of trees where bucking is done without time consuming careful scaling.

In this form of bucking, time required for the whole working cycle is very sensitive to various sizes of the trunk.

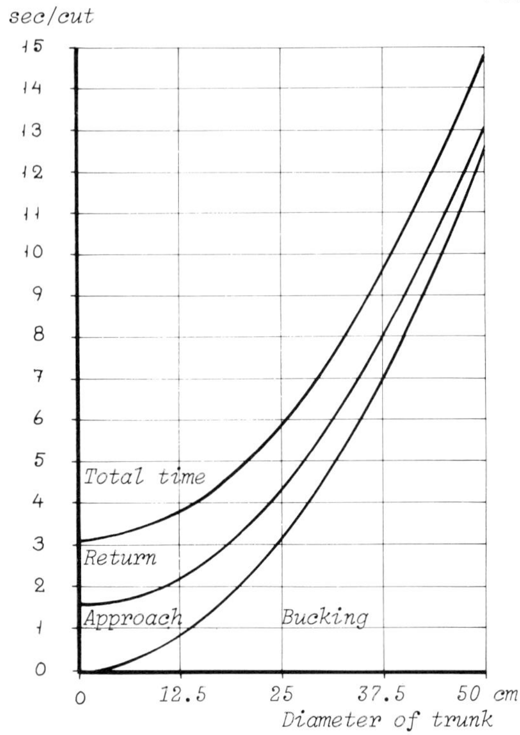

FIGURE 101. Graph showing bucking time for a chain saw cutting trunks of various sizes. Mobile unit (Arnelo and Banner, 1967).

Moving bucking equipment

The moving bucking equipment e.g. hydraulic shears that buck while moving should give such a short time that the whole work cycle can be completed during the time required to feed in a trunk of minimum length. The minimum time of the work cycle is then determined by the rate of feed and the shortest distance between two points of bucking, which for normal standard length of pulpwood is 3 m.

Design and power requirements of the shears depend on the largest size of the trunks.

Shearing (clipping) of pulpwood does not constitute any major technical problem. However, shearing of sawlogs by means of techniques known today is causing too much wood damage to the assortments which give the forest owner the highest revenue.

Interrupted or continuous sequence of bucking

Scaling and bucking at the various points of the trunk can be carried out in an interrupted or continuous sequence, i.e. they can be more or less separated in time.

When the sequence is interrupted, a proper balance between the times of scaling and bucking is required.

When sequence is continuous the requirement for balance is replaced with a requirements for minimum time. The latter condition often applies to processing of whole trees.

Tree characteristics of importance for scaling such as volume, quality and assortments, are difficult to evaluate for limby trees and removal of whole tree trunks impairs the mobility of the bucking depot. When a high mobility is required, the method with automatic scaling and a moving bucking unit is given priority.

An interrupted sequence is easier to utilize in establishments for long logs and trunks of tree length where the requirements for mobility are less stringent, other technical solutions being possible.

At depots for trunks of tree length where there are several bucking units, the operation of the units is balanced. A secondary unit is then able to process the number of long logs delivered by the previous bucking unit.

Bucking of single or several trees

Another principle difference between methods of bucking is a matter of quantity. Bucking can be done either by taking a tree or a trunk individually, or by taking two or more trees or trunks simultaneously, or bunches of trees (slashing).

When big trees and trunks are bucked, the individual processing dominates. The method with bucking of bunches is preferred when the trees and trunks are small. The latter method of bucking is gaining ground particularly when small trees are being harvested. This method will be discussed further in a subsequent chapter.

When the tree part method is used in thinning operations, the grapple saw should be designed in small sizes and with light weight in order to facilitate the movements in the limited space available. The engine powering the chain should be placed adjacent to the grapple which will keep the whole assembly hanging in vertical position.

It is important that the grapple saw has a high chain speed for high productivity. If infeed to the saw is done with a constant pressure, the removal of chips by the saw chain is optimized, the rate of infeed varying with length of cut in the timber.

FIGURE 102. Grapple saw used in the tree part method (Skogen, 1983).

Manual and motor-manual bucking

Bucking at the stump

To some extent bucking of felled trees at the stumps is done by means of one-man saws or bow saws. However, this work is predominantly done by means of chain saws, often in ombination with delimbing and measurements (scaling). In the latter case several different methods are being used, of which a method with special equipment is the most popular one.

The method simplifies the sequence of work because all tools required are conveniently at hand. This arrangement eliminates the time for idle walk and change of tools which is inherent in the conventional method of bucking.

A quick and well done bucking is based on a correct judgement of the position of a tree and situations which cause bending. This judgement determines the choice of a correct method of bucking from above and from below.

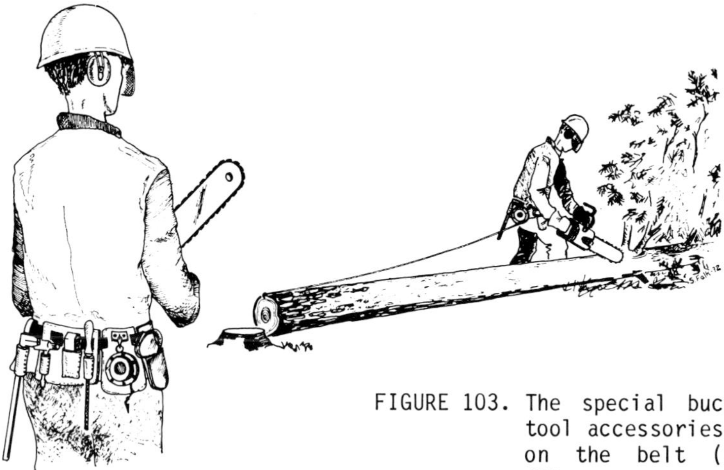

FIGURE 103. The special bucking method and tool accessories. Note equipment on the belt (caliper, wedge, file, tape, screw driver, lifting hook).

When very big and valuable assortments are bucked special work procedures are applied for e.g. situations with tension vertically or laterally.

Bucking at various tension conditions

Tension from bottom occurs when the tree lies over a solid object or when the tree is firmly supported in one end and has an over-hang at the other end. Bucking starts where the saw bar is liable to jam first i.e. the low side of the trunk. Bucking is completed from the top side, opening a cut into the wood by means of the saw (see Figure 104). Numbers in Figure 104 give the order in which bucking should be done (Conway, 1982).

Tension from the sides often occurs (Figure 105) when the trunk is bent sideways. When the bucking is done, the end swings in the direction indicated, far side of the tree being under compression while the near side is under tension.

A drop tension occurs when bucking is made at an angle that will allow the end to drop, preventing jamming (see Figure 106). Bucking is done by means of the technique used for bucking at top tension.

It is particularly important to adapt to the various kinds of tension when big, valuable timber is being bucked.

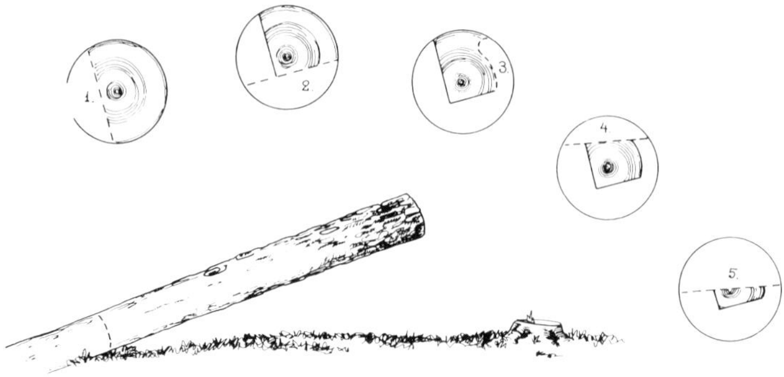

FIGURE 104. Bucking procedure for very big and valuable assortments at bottom tension.

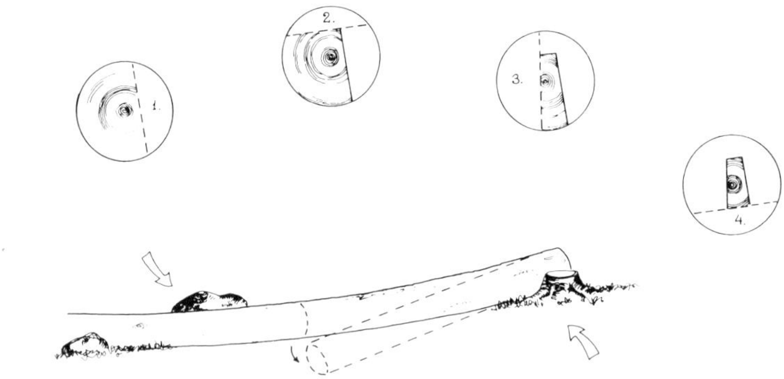

FIGURE 105. Bucking procedure at tension from the side.

FIGURE 106. Bucking above a depression.

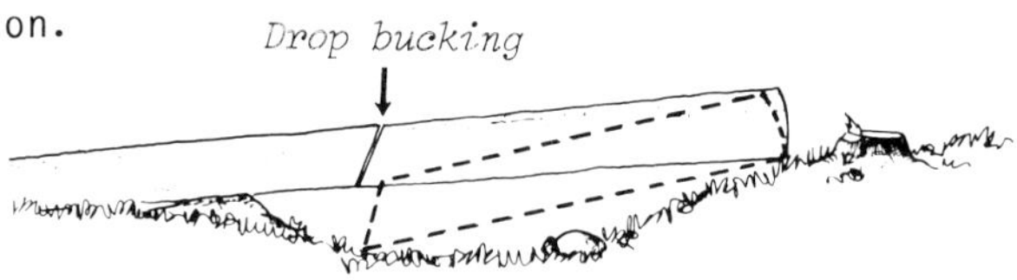

Bucking at strip roads

When delimbed trunks are placed in rows or piles along strip roads e.g. after a machine operation, bucking by means of chain saw is usually done in a conventional way for assortments which are then stacked for forwarding.

Combined motor-manual and mechanized bucking or bunches left after a machine has been tried by means of a grapple saw. The saw has been mounted on a regular forwarder in order to enable bucking of the sawtimber portion of the trunks. The method has been called grapple saw forwarding and it is discussed further in a chapter on mechanized bucking. The subsequent bucking of pulpwood is predominantly done manually by means of chain saw because of difficulties encountered when measuring 3-m bolts. When trunks are processed, the top ot the tree is cut at a size that is assumed to hold up to skidding by winch. Skidding by winch however, is increasingly being replaced with skidding by means of clam bunks.

Bucking at landings

Bucking at landings is carried out by means of chain saws and it is often followed by manual bunching of pulpwood into shorings or piles. However, bunching is increasingly being done by means of fork loaders or grapple loaders.

The landing should be sufficiently large to accommodate the bucking operation with its transport of timber to and from the landing. If fork loaders are used for the handling of timber, the area of the landing should be at least 2000 m^2. A minimum of two shorings for processing should be built on the landing in order to enable unloading of trunks, bucking, bunching and truck loading simultaneously and independently of each other.

The build-up of a "buffer stock" of trunks will facilitate a continuous operation.

Mechanized bucking

Mechanized bucking can be carried out by means of chain saws, circular saws and clipping-shearing tools of various designs.

Mechanized bucking at the stump

Mechanized bucking at the stump is done by means of equipment which cuts off the trunks into bolts of standard length by means of a single stroke hydraulic knife. The bolts are then dropped into a bunch collector. Feeding is arranged by means of a hydraulic cylinder which is pushing a claw like attachement pierced into the trunk. The length of feed corresponds to the standard length of the bolt. This length may vary since processing of random length logs can be carried out according to the same principle.

When a fixed length is fed for bucking, the feeding mechanism returns automatically in order to take a new hold on the trunk. Feeding of the butt end and delimbing of the top proceed at the same rate.

Mechanized bucking at strip roads

Grapple saw. The grapple saw (Figure 107) is used for bucking of sawtimber from bunched trunks along the strip roads before conventional bucking by means of chain saw is being done for the remaining pulpwood sections.

FIGURE 107. Grapple saw.

The saw, which is mounted on a hydraulic grapple with a cross-section area of 0.35 m^2 and operated by means of a hydraulic engine, has been tested during field trials i.a. on a crane with a reach of 5.7 m. The loading machine used is a forwarder. Grapple loaders are used more recently.

The best view of the bucking operation when this unit is used is obtained when travelling towards the top ends of the bunches.

Method of bucking by means of the grapple saw

The machine is positioned so that the trunks can be pulled from the bunch, each trunk being grabbed as near to the first point of bucking as possible. The operator then determines more exactly where bucking should be done. He moves the grapple to the point of bucking and cuts the trunk. The remaining part of the trunk is forwarded for bucking of sawlog no. 2 and so on. The pulpwood portion is put aside into an even pile in order to facilitate the subsequent motor-manual bucking into bolts, usually done by machine at truck road. During the process of bucking the sawlogs are loaded onto the forwarder at suitable intervals. When the forwarder is loaded, it moves to a landing at the truck road.

Method of bucking by means of a very rapid chain saw

Machine manufacturers have tried to replace the space demanding and risky circular saws used for bucking in processors and harvesters with built-in and safe chain saws. The problem in cutting suspended parts of trunks by means of chain saws is cracking in the end surface. One attempt at solving this problem has been made by increasing the chain speed to at least 45 m/s. It was possible to increase the original chain speed from 20 m/s to 47 m/s by means of a roll-top bar. This speed gives a cutting capacity of 1200 cm^2/s which is within the safety margin for cracking. At this chain speed it is very important that maintenance of the bar and in particular the saw chain is strict. The chain must be exchanged every day. Filing is necessary for each dulled tooth. Usually it is sufficient with 2-3 strokes to sharpen a tooth. Bucking without cracking should always be required.

FIGURE 108. Bucking by means of chain saw on harvester, 1983. The log is "hanging" horizontally in the air after being cut.

Processor A

The processor is a mobile machine for processing primarily at final harvest operations. The machine delimbs, bucks and deposits the timber in bunches.

In principle the machine applies the same method of work as that of a previous processor. For processing at landings a model shown in Figure 90 is feasible.

Method of felling

In each position all trees within reach of the crane are processed. Each tree is grabbed approximately 3 m from the butt end and brought into the machine for processing.

The trees are felled away from the strip roads (or the paths of driving) in a direction approximately at 90° angle (± 15°) to the road. The trees standing far away from the road in a felling swath of 40-70 m are felled first. In dense stands a crowded felling into the untouched stand can be avoided if felling starts in the more open parts of the stand.

Since the requirement concerning directed felling is strong, two alternative directions of felling should be considered in view of the predomi-

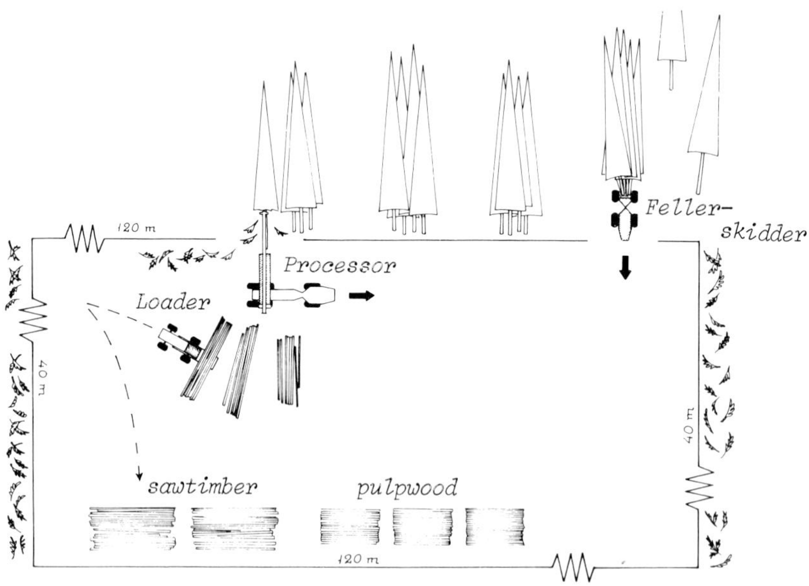

FIGURE 109. Feller-skidder and processor at a landing

nant wind directions. Simultaneously, attempts should be made to lay out the strip roads parallel to the contours in order to facilitate the bunching of pulpwood.

Various partial operations

Work can be divided into the partial operations of infeed, processing, sorting and moving to a new position. The machine can also be used for processing in the tree method.

The machine is built on a chassi with a rear bogie. The parallel feeding boom with grapple for holding and rough delimbing is mounted together with the processing unit with a pivot on the rear carier of the base machine.

The feeding boom has a reach of 12 m from the pivot to the grapple.

When bucking is done by means of a circular saw with a high peripheral velocity or chain saw, sawlogs drop via a sorting deck directly to the ground while pulpwood is collected in a pocket with a capacity of 2 m^3.

Bucking of sawlogs is done by means of manual release while bucking of pulpwood is automatic.

Terrain travel

It is obvious that the terrain conditions will influence the performance of this machine. However, influence is not so great as expected in view of the size of the machine due to the ability of the carrier to travel in terrain.

Processor B

The processor is a delimber-bucker-buncher. In its present design the machine is primarily intended for processing of trees bunched along the strip roads in a final harvest operation. Presently, the machine is equipped with a knuckle boom but work is underway on the development of a supplementary crane with a longer reach for processing of trees felled parallel to each other in the stand.

Since the machine can also be used for processing of bunches of trees at landings or at truck roads, it can be integrated with various systems of harvesting.

A system suitable for the machine is the combination feller-buncher-processor on strip roads + forwarder. This system is shown in Figure 110.

Work procedure

A feller-buncher operates along the border of the stand depositing bunches in strings at an interval of approximately 4-5 m depending on density of felling and distance between the rows of bunches (40-70 m).

The processor, working in a direction shown in Figure 110, is positioned in relation to the bunches so that the trees can be grabbed easily and placed on the feeding bunk without retake. The feeding bunk, which can hold several trees, works as a buffer storage between infeed and processing operations. The latter process, therefore, can be carried out rather continuously.

After the desired length of the timber has been set, the delimbing, bunching and topping processes are entirely automatic. The sawlogs fall directly to the ground while the pulpwood bolts are collected in pockets of which there are two in case a separation of two assortments is desired.

To keep the pulpwood in piles, the pockets are lowered to the ground and emptied in suitable places beside or behind the machine.

FIGURE 110. Processor B integrated with a harvesting system where it bucks and delimbs at strip roads.

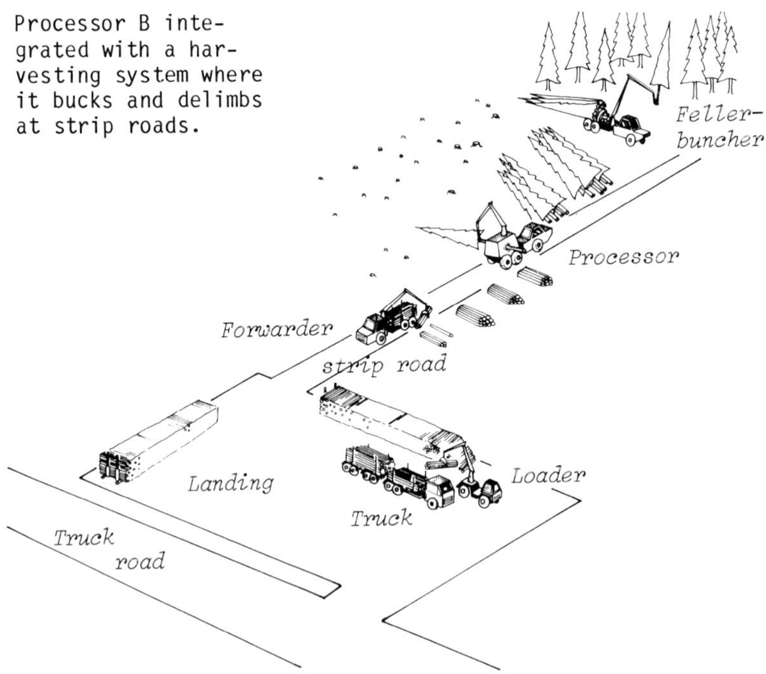

Design of the machine

The processing unit with the operator's cabin is placed on a turning ring above the rear carriage. On top of the cabin is mounted a knuckle boom for loading with a reach of 8.5 m.

Processing of timber

After bunching the trees are placed on the infeed bunk which can be raised or lowered. It is equipped with a feeding crest that is designed to bring forward one large tree separately or several small trees simultaneously for processing. Meanwhile, the other trees on the infeed bunk are kept in place which provides for the necessary separation of the trees from the bunch. Due to this feeding procedure, production becomes less dependent on diameter of the trees.

Delimbing is done by means of two bands anchored partly in the frame of the processing unit and partly in two overlapping hydraulically maneuvered braces. This mounting gives a good fit to the trunk(s). Feeding is done by

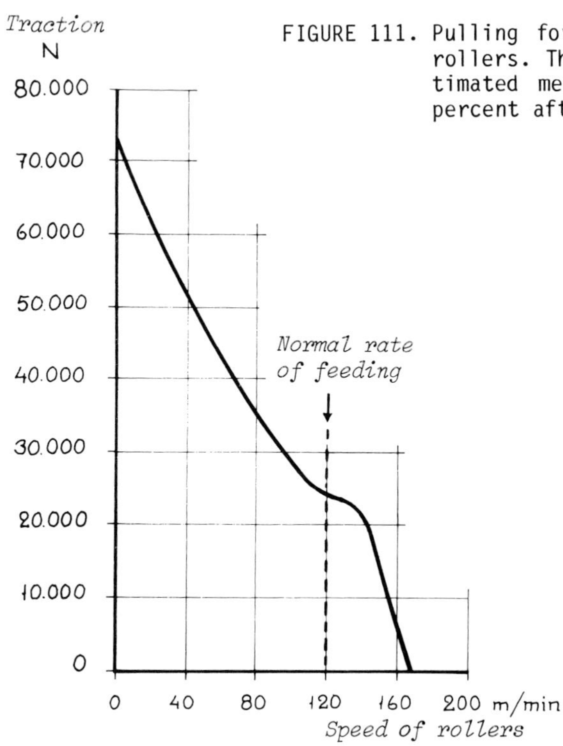

FIGURE 111. Pulling force and speed of the feed rollers. The graph is based on an estimated mechanical efficiency of 85 percent after a torque converter.

means of two cylindric spike or rubber rollers which are powered by the processor engine through a converter, power shift gear box and separate gears. This arrangement provides for an adjustment of the pull and speed of the feed rollers to the actual resistance to delimbing.

At normal rate of feeding, 120 m/min, pull is 24 000 N (2400 kp) and at a very low rate of feeding it can amount to 72 000 N (7200 kp). See Figure 111. Pressure on the feed rollers varies with pull utilized. The rate of feeding increases rapidly at reduced diameter of the trees which can be recognized ocularly.

Processor B is a harvesting machine with automatic, variable rate of feeding according to principles described above. This design feature further reduces the dependence of output on sizes of trees.

Bucking is done automatically by means of a hydraulic circular saw or chain saw which can cut sawtimber into logs of random length by 0.6-m classes between 3.6 m and 6.0 m and pulpwood into 3-m length or multiples of 0.6 m.

The top is cut off at the delimbing mechanism by impulse from a diameter sensor. Out-feed and sorting into pockets are also automatic.

The cabin has been built for two operators, taking into consideration ergonomic and work inducing requirements. The operators are placed so that they have the processing mechanism in front.

Mechanized bucking at truck roads or industrial landings

Bucking of partial trunks

Partial trunks, which remain for pulpwood after the sawlogs have been cut off by means of grapple saw at the strip roads, can be bucked in bunches at truck roads or at industrial landings.

A forwarder transports the partial logs of pulpwood to a landing. The trunks are then collected into feasible bunches that can be bucked into standard length by means of a circular saw. The circular saw with a pocket for timber is powered by the mobile base machine. The base machine is also equipped with a knuckle boom and grapple for putting the bunch into a pocket or for piling of the bucked pulpwood.

A bucking unit of this type can produce approximately 30-40 m^3 per hour of efficient work, primarily depending on the number of assortments required.

Bucking of trunks in the tree length method

Bucking units. Interest in mechanized processing of trunks at truck roads or landings has been great ever since the tree length method was introduced.

Several types of bucking units have been designed. One bucking unit has recently been combined with a delimber-buncher.

Mobile bucking unit. The unit is mobile, the base machine being a three-axle truck with a 110 hp engine. A hydraulic grapple and a bucking unit with circular saw or chain saw, feeding and support rollers, push-off and timber pockets are included in the processing machine. The machine can be run by one or two operators.

All handling of the timber, infeed, outfeed and sorting is carried out by means of a knuckle boom equipped with a grapple.

Bucking of trunks in the tree method

Processor C

Processor C is a Canadian machine for delimbing, debarking, bucking and bunching of pulpwood at landings or at truck roads (Horncastel, 1965).

Trees suitable for pulpwood are transported by means of a skidder to a place for processing where they are put in an even row, butt end pointing forward and at 90° angle to the travelling direction of the machine which is usually along the road side during the processing operation (Figure 93).

A telescoping boom with grapple brings the tree into the machine where it is delimbed, debarked and bucked into bolts of standard length in a compact processing unit. The bolts are collected in a timber pocket. When the pocket is full it is emptied through an opening in the bottom.

Immediately in front of the feed rollers is placed the delimbing mechanism described in a previous section on delimbing. Topping is done by the same mechanism. Just behind the feed rollers there is a debarking unit, with a maximum output of 37 kW (50 hp). The tools of the debarking unit rotate in the same direction as the delimbing rotor. After debarking, the trunks enter between two hydraulic knives, so-called double acting shears, which are opened and closed in less than three quarters of a second.

During the moment of cutting the shears follow the trunk and work as a moving bucking saw. Thus, feeding need not be stopped for the bucking operation.

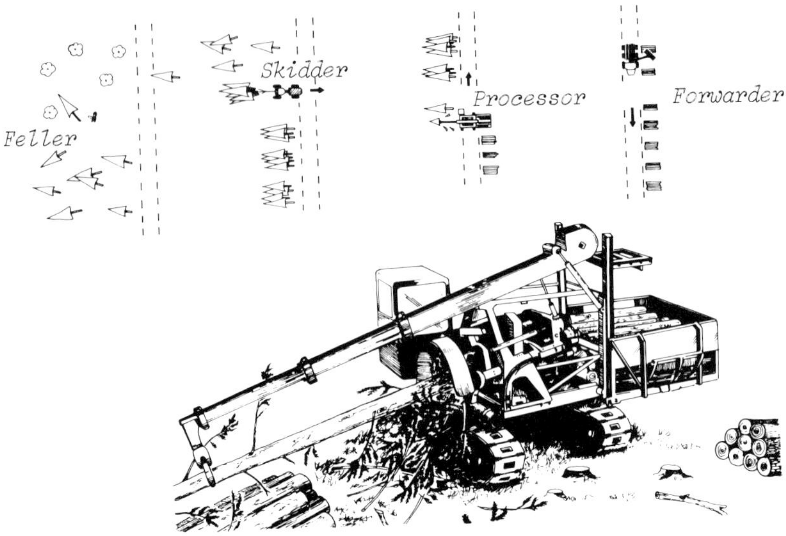

FIGURE 112. Processor delimbs, bucks and bunches.

Output and costs

The machine is able to process approximately two 12-m trees per minute. However, the production is strongly dependent on the sizes of the trees when this method of processing one tree at a time is applied. Variation in the sizes of the trees is more decisive for the output than the variation in the rate of feeding (cf. Figure 113).

Interest in this machine cooled when debarking in the forests declined. Simultaneously it appeared that the debarking process influenced strongly the rate of feeding. Moreover, the problem with bunching after the out-feed has not found an acceptable solution for all conditions.

The machine has a good integration of various partial operations and because of the efficient moving bucking mechanism it is of great interest from the point of technology (methods) and deserves a presentation in this context. Still, the trend in using mobile machines and units for a complete processing at landings is declining.

Mechanized processing of tree length trunks at industry or terminal

Several combinations of machines for industrialized bucking, sorting, debarking and chipping of timber have beed designed. A modern combination is presented here.

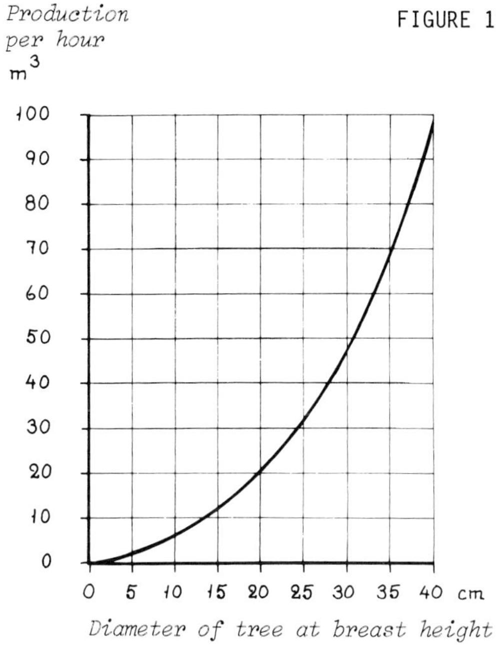

FIGURE 113. Production per hour achieved by means of a processor. Rate of feeding is approximately 45 m per minute.

Main components of the establishment

The combination consists of the following main components: Infeed deck, maneuvering bridge with control centre, measurement unit with bucking saw, conveyors with pockets for unloading of sawlogs, pulpwood and special assortments. It may also include debarking machines, chipper with cyclone, chip screen and chip conveyor to bins.

A combination of this kind is usually run by a scaling-machine operator and a truck driver for infeed of trunks.

Bucking - scaling

Programming of scaling and sorting is done by means of push-buttons on a control panel and data can be stored for a maximum of four trunks at a time.

The automatic equipment allows scaling and bucking of sawlogs and special assortments in a number (e.g. 12) of length classes (feet and/or metres), and pulpwood in standard length (most frequently 3 m) which can be sorted into six groups (three before and three after debarking). The pulpwood portion can also be transferred directly to the debarking machine and chippers.

Certainly, capacity of the machines largely depends on the requirements with respect to bucking (scaling).

The average capacity of a combination of this kind may be 300 m^3 per shift if the average size of the trunks is 0.5 m^3. In one-shift operation this output corresponds to approximately 65 000 m^3 per year and in a two-shift operation 120 000 m^3 per year.

Mechanized processing at mobile and semi-stationary establishments

An analysis of various methods of processing can produce a required representative amount of data on trees for i.a. a theoretical calculation of yield in various diameter classes (Arnelo and Banner, 1967).

After a distinction of various types of processing from trees to log length through trunks and multiple log length, potential principle solutions are investigated for various machine components combined into establishments. The point of special interest when comparing different establishments is primarily the cost of a given process and the value produced within a complete harvesting system including its industrial interface.

Volume of work is the difference between input i.e. the product that is supplied to the establishment (tree, whole trunks or multi-log piece of trunks) and the output i.e. the product that is delivered by the establishment such as trunks, multi-log pieces of trunks and logs.

Multi-log pieces of trunks are partially bucked timber containing one assortment only and cut off at the minimum diameter limit.

Partial operations in the analysis

The following operations are of interest in the analyses mentioned above:

- Infeed	Dependent and decisive for output
- Delimbing	-"-
- Scaling, bucking	-"-
- Sorting	-"-
- Debarking	Independent and indecisive for output
- Chipping	-"-
- Handling	-"-

Table 6. Various types of processing establishments

No.	Input	Output	Components
1	Trees	Whole trunks (Saw-timber and pulpwood)	Delimber
2	Trees	Multiple sawlogs Pulpwood	Delimber and moving bucking saw
3	Trees	Sawlogs Pulpwood	Delimber and two fixed bucking saws for automatic and manual scaling
4	Trees	Sawlogs Pulpwood	Delimber and two bucking saws in fixed positions for manual and automatic scaling
5	Trees	Sawlogs Pulpwood	Delimber, one bucking saw in fixed position and one moving bucking saw.
6	Small trees	Pulpwood	Delimber and one moving bucking saw
7	Whole trunks	Sawlogs Pulpwood	Storage deck with infeed for single or several trunks, moving bucking saw for pulpwood, bucking saw in fixed position for sawlogs.
8	Whole trunks (small)	Pulpwood	Storage deck with infeed for single trunks, moving bucking saw.
9	Multiple sawlogs	Sawlogs	Storage deck with infeed for single multiple sawlogs, bucking saw in fixed position.

Table 6 shows how the operations can be combined in processing establishments at various locations between the stumps and the mills.

The number of options is very large, at least 42 different establishments with 28 different machine combinations are possible, e.g.

- For processing of whole trees there are 12 different establishments and 18 machine combinations possible
- For processing of tree length trunks there are 27 different establishments and 4 machine combinations
- For processing of multiple logs there are 3 different establishments and 6 machine combinations

This wide range of combinations depends i.a. on the varying distribution of trunk sizes which determines the degree of processing necessary for a given volume proportion of sawtimber and pulpwood. It also depends on the availability of various types of machines and machine components with different modes of operation, capacity, precision and mobility. (Cf. Platzer, 1970, Platzer, Wipperman, 1972 and Kaminski, 1981).

Infeed

The infeed unit is an important component in the processing establishment particularly in the case of whole trees. Size of the unit is determined by the weight of the largest trees. Infeed can be arranged either for single trees directly into a delimbing machine or for bunches of trees or trunks placed on a storage deck.

Whole trunks and multiple logs can be fed via a storage deck with a length-wise infeed unit. A high infeed capacity is required for a complete utilization of the whole establishment when the scaling and bucking processes are quick due to automatic or semi-automatic components.

Delimbing

Mechanized delimbing with longitudinal infeed of single trees is applied in the various types of establishments. Because of requirements for mobility and a rational coordination with the bucking process, delimbing of several trees at a time is making slow progress. It is important that the size of the delimbing machine is adapted to the sizes of trees and limbs, not least from the point of machine cost.

Scaling and bucking

The scaling process may be considered from two viewpoints:

- From a technical viewpoint scaling is often lowering the output, particularly when it has to be careful and tedious
- From an economic viewpoint scaling determines the value of output, particularly for sawtimber and other assortments sensitive to scaling.

At simplified scaling the loss of value may soon exceed the gain of rationalized processing, perhaps even jeopardizing the whole mechanization effort. Several alternative principles of scaling can be applied in a processing establishment:

1. Manual scaling with manually operated bucking
2. Semi-automatic scaling with entirely or partially manual scaling and automatically operated bucking
3. Entirely automatic scaling with automatically operated bucking.

Choice of scaling alternative primarily depends on the sawtimber quality required. Scaling of sawtimber in turn is influenced by the price of sawn goods (lumber).

Sorting

Sorting is a partial operation done by means of feed rollers, push-offs, sorting pockets with control mechanisms for two or several assortments. In normal types of establishments the cost of sorting increases with the number of assortments (Figure 114).

Handling of timber in a processing establishment

The handling of material in a processing establishment is a major cost item, particularly when tree length trunks and multiple logs are to be processed. Handling of timber can be done by means of various types of loading machines.

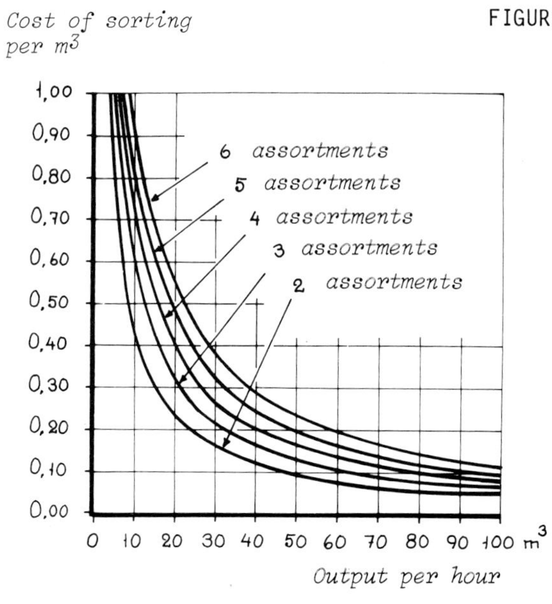

FIGURE 114. Graph showing the costs of sorting (per m^3) at various output (Arnelo and Banner, 1967).

Trends in bucking

Increased mechanized bucking

Increased mechanization of tree harvesting operations involves bucking. Mechanized delimbing in machine units of the type delimber-buncher has increased steadily in recent years, particularly at final harvest operations. This method has generally been combined with a motor-manual bucking by means of chain saw. This way of bucking is now being replaced gradually with separate mechanized bucking by means of grapple saws or mobile bucking units.

Integration of delimbing, bucking and bunching by means of processors has increased steadily. It has been considered consequential also to combine mechanized delimbing with machine bucking in the same unit (Figure 115).

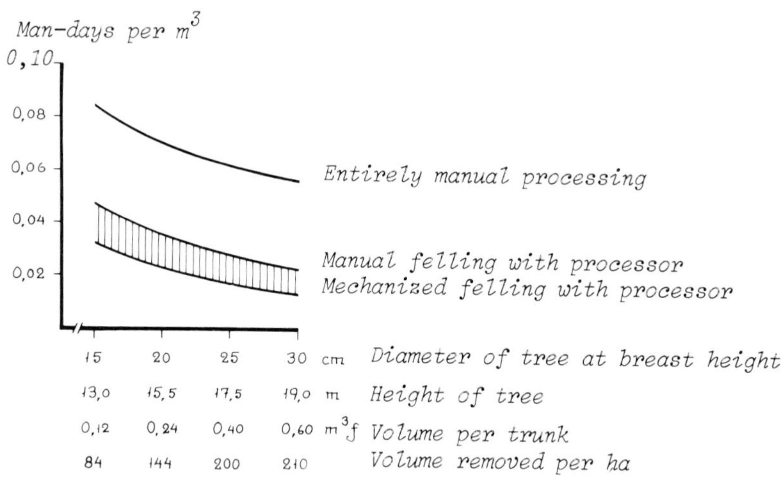

FIGURE 115. Influence of tree size on the labour input (man-days per m^3) at entirely manual processing, at manual felling with processor, and at mechanized felling with processor.

Scaling has been a problem at mechanized bucking. A poorly done scaling of timber easily incurs great losses in value, particularly when big and valuable trunks are processed. These losses can surpass the savings in harvesting costs that can be achieved by mechanized operation.

When a delimber-buncher is used, the important scaling procedure may be put into any point of the harvesting sequence. Thus, the choice of harvesting method is influenced by the care of scaling desired for the quantity of timber to be cut.

Automatic scaling and bucking

Application of electronics

Techniques used at automation of manufacturing processes in general can also be applied to mechanized processing of timber and primarily for automation of scaling, bucking and sorting.

The first establishment with programmed bucking and sorting in Sweden was built in 1964. The operator could select on a control panel a scaling program taking into consideration the length of logs desired. The control system was based on relays, a technique common at the time.

A first step towards automatic scaling by means of new techniques was taken a couple of years later when a prototype equipment was tested.

Since the relays have now been replaced with integrated electronic circuits of great sophistication, compact and insensitive to environment (vibrations, temperature, moisture etc), it has also been possible to build in equipment and instruments for measurements and controls in mobile machines.

Photo-cells

Automatic measurements of the diameter of sawtimber can be achieved by means of an automatic scaler with photo-cells attached to the bucking unit.

The log is passing the measuring device according to Figure 97. In each of the two arms A_1 and A_2 is enclosed a band with a crosswise slot along a row of small light bulbs. The band moves with a speed of 30 m per sec. Light from the lamps, which is successively passed through the slot, is received by a photo-cell placed opposite each arm.

Each photo-cell generates ten pulse series per second which are alternately fed into a tally with two registers. A log which is passed between the two arms intercepts the light from a number of lamps and, hence, reduces the number of pulses in each series that reaches the photo-cells. The number of pulses lacking in the photo-cells is a measure of the diameter of the log.

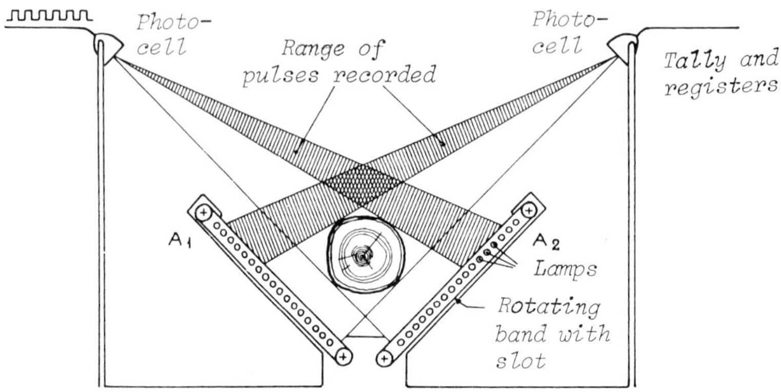

FIGURE 116. Automatic scaler for diameter measurements.

Debarking

Debarking is the work involved in removing bark from the trunks. Degree of debarking is used to express the extent of mantle surface that has been debarked, e.g. the entire trunk, half the trunk, debarking in spots or strips.

Purposes of debarking

Debarking in the forest

Debarking in the forest is primarily done in order:

- to promote a relatively quick drying of the timber and, hence, reduced storage defects
- to reduce weight and to some extent volume of timber, particularly for transports on land

Debarking at the industry

Debarking at the industry is done for the following main reasons:

- to obtain a good quality of the finished product, primarily pulp and sawn goods (lumber)

- to promote a reduction of the amount of chemicals and bleaching compounds at the pulp manufacturing process and to obtain a better utilization of the pulpwood digesters.

Choice of location for debarking

Choosing the forest or the industry for debarking is a matter of transportation and integration of forest operations with industrial processing.

Timber is increasingly being debarked at the industry. The reason for the rapid increase in industrial debarking is discussed in the end of the chapter on debarking. Although debarking in the forest has declined, means and techniques for this type of operation will be discussed partly because it is still prevalent, partly because it can be a necessary practice in several special cases, e.g. in certain storage situations.

Some physiological features of bark

The anatomy of bark in various species can be studied separately under the subject of forest botany (Cf. Knigge u. Schulz, 1966).

Various layers of bark

Bark is composed of three layers: outer bark (cork), inner bark or bast, and cambium. These layers can be classified with respect to anatomic structure or physiology. Volume and weight of bark in relation to volume and weight of the tree has been described and statistics are given in a previous chapter.

Cambium

Cambium is a distinct layer between the wood and the outer layers of bark. The fibers of the cambium in coniferous trees have a length of 1-2 mm and a width of 0.03 mm. Growth of the cambium fibers by means of tangential walls is producing bark fibers on the outside and wood fibers on the inside.

Inner bark

Inner bark is the layer extending from the sieve fibers and the sieve tubes produced by the cambium to the last formed lignified parenchymatic fibers.

The inner bark varies in thickness between 3 mm and 10 mm. In pine and larch it is 1-3 mm thick, in spruce and beech about 15 mm.

Transporting nutrient solutions from the crown to the roots, the inner bark carries a large amount of water. When the water in the inner bark is freezing, debarking is considerably more difficult. Cellulose is a major component of the inner bark, e.g. approximately 25 percent in pine.

Outer bark

Outer bark is the layer extending from the last formed parenchymatic fibers to the surface of the trunk. In its interior parts the outer bark consists of a layer of parenchymatic fibers while the outer parts consist of a layer of cork with dead fibers.

The function of the outer bark is to protect the tree from mechanical and chemical influences and from fungi. The bark also insulates the wood from extreme temperature fluctuations. There are trees with a bark thickness of approximately 50 cm (cf. Sequoia) which is good protection from e.g. forest fires.

The proportions of inner and outer bark determine the choice of method and means of debarking. Lignin is the major component of the outer bark, in pine approximately 44 percent.

There are two characteristics that determine the difficulties of debarking, viz. the configuration and smoothness of the bark surface, and cohesion between bark and wood.

The extent of bark surface is related to the diameter and the length of the trunk. The configuration and smoothness of the surface depends i.a. on the occurrence of crooks, stubs of limbs, bumps, forks etc. All these features influence the work of debarking.

Cohesion between bark and wood

The greatest effort in debarking is spent on dealing with the cohesion between the bark and the wood. The force needed for debarking depends on the cohesion between bark and wood and on the bark thickness.

An example of the cohesion in spruce (Picea mariana and Picea rubens) in eastern Canada is shown in Figure 117 (Berlyn, 1965).

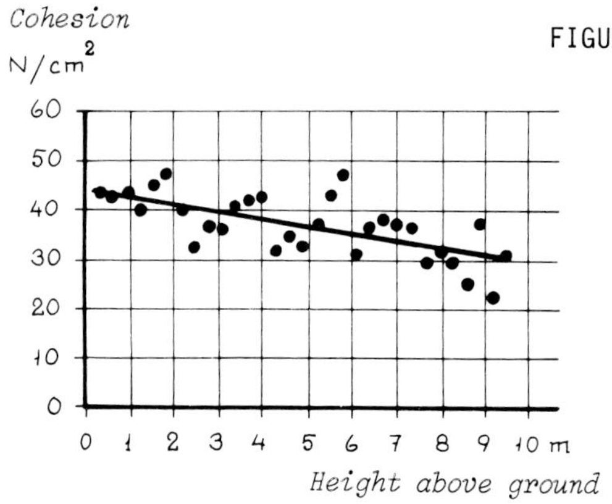

FIGURE 117. Cohesion between bark and wood at various height above the ground. (Picea mariana and Picea rubens).

The values apply to a tree with a diameter of 22.9 cm at breast height. Bark facing south. Summer.

Cohesion between bark and wood varies between species and seasons. The equipment used in the Canadian tests above failed to record clearly the influence of bark thickness. Some Russian investigations (Figure 118) report on the influence of temperature on the cohesion between bark and wood in pine and spruce (Pinus silvestris and Picea excelsa).

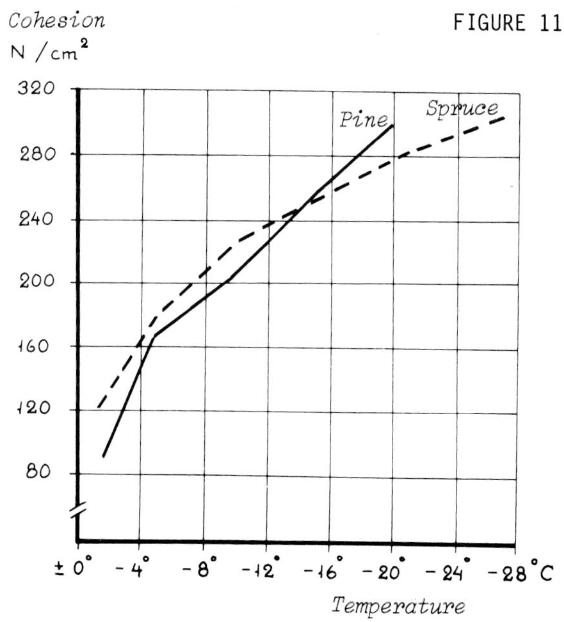

FIGURE 118. Cohesion between bark and wood in pine and spruce at various temperatures (Voronitsin and Vorobyev, 1965).

According to Russian investigations cohesion amounts to approximately 250 N (25 kp) per cm^2 for pine and spruce at -14°C and the values continue to increase at declining temperature.

Measurements carried out in Sweden show that the bark is semi-resilient at a temperature between -2°C and -8°C before it becomes frozen solid at a temperature between -8°C and -9°C. Once the bark is frozen solid its characteristics are constant at further decline in temperature.

To deal with the cohesive forces reported above, debarking is done by means of various mechanical, hydraulic and chemical methods. Electric equipment and injections with steam or explosive gases have also been tried.

Tools and means of debarking

Manual debarking in the forest

Manual debarking by means of a spud was the predominant method of debarking in the Swedish forests up to the mid 1950's when machines were introduced to an increasing extent.

In 1960 about 40 percent of all timber was still debarked manually. The proportion of manually debarked timber decreased rapidly to about 10 percent in 1965. Manual debarking is now done to a very minor extent.

Manual debarking is physiologically very strenuous work, particularly in winter when bark is frozen. About 50-60 percent of the total time required for the manual processing of a tree is then spent on debarking.

Debarking can also be done by means of a knife or an axe. The latter tool is used for very thick bark, e.g. on the butt end of a big tree.

Manual work required for debarking or cutting off bark from tree trunks is influenced by:

- size of the trunk which determines the number, width and thickness of bark strips at various degrees of debarking
- physical characteristics of the bark depending on i.a. species, seasonal temperature, age and size of the tree
- technical design and sharpness of the debarking tools.

Motor-manual debarking

To eliminate the physically very strenuous and time consuming manual debarking, it was of great interest to develop portable machine tools. Machines developed for this purpose in the beginning of the 1950's are now of limited interest. The types that were marketed did not reduce the work load nor did they increase productivity. In addition the machines worked with severe vibrations and high noise levels. The machines could not compete with the specialized debarking machines that were being developed.

Mechanized debarking

A large number of machines have been designed for mechanized debarking at strip roads, truck roads or landings. Three types can be distinguished with respect to the technical means used for debarking: debarkers with knives, cutters and rings (rotors).

Debarkers with knives

The debarking machines with knives usually have 3-5 knives radially mounted on a circular, rotating disc.

During the debarking operation the rough trunk is pressed under rotation against the rotating disc. The knives remove the bark and, unfortunately, a certain amount of wood, up to 10-20 percent.

Debarkers with cutters

The mechanism of this type of debarkers consists of routers or cutter heads that rout or cut off the bark in various ways. The debarkers can be used for stationary operation or, if mounted on a tractor, in mobile units.

Debarkers with rings or rotors

The debarker with rings is the type of machine which best meets the requirements for high output. The debarking mechanism consists of 5-8 tools rotating around the trunk. The tools are held by means of springs against the trunk, gliding on the surface of the wood along a helical coarse. The tangential pressure of the tools give rise to pushing forces sufficient to crack the cambium layer between bark and wood. The debarking tools are blunt for the purpose of avoiding damage to the wood.

The tools may have either a combination of cutting and scraping-skewing actions or a work distribution between the tools, every other tool cutting a groove in the bark and the other tools scraping loose a corresponding strip.

Preceding the modern debarking machines with rotors was a debarking machine developed in the beginning of the 1950's. It was stationary and designed for debarking of sawlogs (Figure 119).

About 150 machines of this type had been sold and delivered on the international and domestic markets by 1954. This sales result made the Swedish Andersson machine the most commonly sold debarking machine at the time. It was i.a. tested on a truck load of yellow pine timber from southern U.S.A. shipped to Sweden by air.

The rate of infeed was approximately 30 m/min or about 360 sawlogs/h at an average log length of 5 m.

FIGURE 119. The Andersson machine. It was primarily the design of the debarking tools and the basic principle of operation that put the machine in the forefront.

FIGURE 120. Detail of the debarking machine.

FIGURE 121. Debarking tools with a combined cutting and scraping action

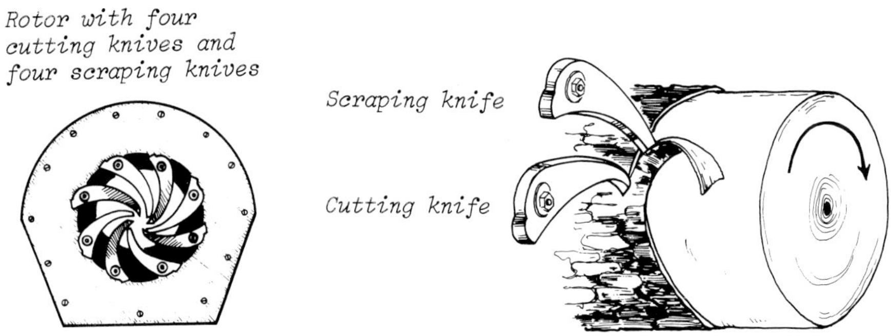

FIGURE 122. Debarking tools with a work distribution (V-K, Valon Kone)

Working principles of a debarking machine

Cambio, following the Andersson machine in 1955, is a very common debarking machine. A large number of sizes and designs for mobile or stationary debarking are manufactured. Representing the group of debarkers with rings, this machine requires a detailed description (Figure 121).

The Cambio machine has become a well-known debarker on the world market on account of its good design and large capacity. It is manufactured in six different sizes for logs ranging in maximum diameter between 35 cm and 108 cm. The machine is made as a complete unit with engines for rotor and feeding mechanisms mounted in the base of the machine. Timber is fed through the machine, one log at a time, by three rotating spike rollers in front of, and by three spike rollers behind the debarking mechanism. When the trunk is moving through the machine, bark is peeled off by means of rotating knives attached to the rotor. The rotor is also equipped with fan blades for removal of the loose pieces of bark.

Pressure of the debarking tools

Pressure of the debarking tools against the trunk is exerted by means of heavy rubber bands. These bands can be stretched, as required because of low temperature and debarking difficulties, by means of a special lift (jack) delivered with the machine. Tension can be observed on a graduated scale. When a certain tension is achieved, the tension disc is locked in position by a simple maneuver.

Procedure of debarking

A frontal, sharp and concave edge of the debarking tools is curved slightly upward. When the trunk is fed by the rollers towards the debarking tools, it first meets the sharp edges that cut into the end of the trunk. Since the tools are rotating, they are forced to glide along their incisions in the end of the trunk until they reach the surface of the bark.The whole movement occurs within a fraction of a second - no slowing down of the trunk can be perceived by eye. Once on top of the bark surface the tools immediately start debarking.

Since the width of the edges in the debarking tools is 2 cm, each revolution of the rotor with its 5 tools will produce a 10 cm peeling of bark along the trunk.

Figure 123 shows how debarking capacity varies in principle at various diameter and rates of feeding.

The following comments can be made after a discussion and interpretation of the curves shown in Figure 123, demonstrating the relationships between debarking output i.e. path of the debarking tools around the trunk, diameter of trunk, bark thickness and volume of debarked timber at various rates of feeding.

- When a 20-cm trunk is debarked at a rate of 100 rpm, a volume of 0.31 m^3 debarked timber is obtained. An approximately equal volume is obtained when a 10-cm trunk is debarked at a rate of 400 rpm (0.32 m^3).
- Work done by the debarking machine on:

  a) the 20-cm trunk is the product of path, here equivalent to 72.8 m, and cohesion between the bark and the wood (in this case 1.3 cm bark thickness for pine of intermediate type),

b) the 10-cm trunk is the product of path (125.6 m) and cohesion between bark and wood (0.65 cm bark thickness)

- It should be possible to alleviate the strong influence of trunk diameter on output by varying the rate of feeding, because the cohesive resistance at reduced bark thickness in small tree trunks should be lowering and stabilizing the power requirements in spite of increased rate of feeding.

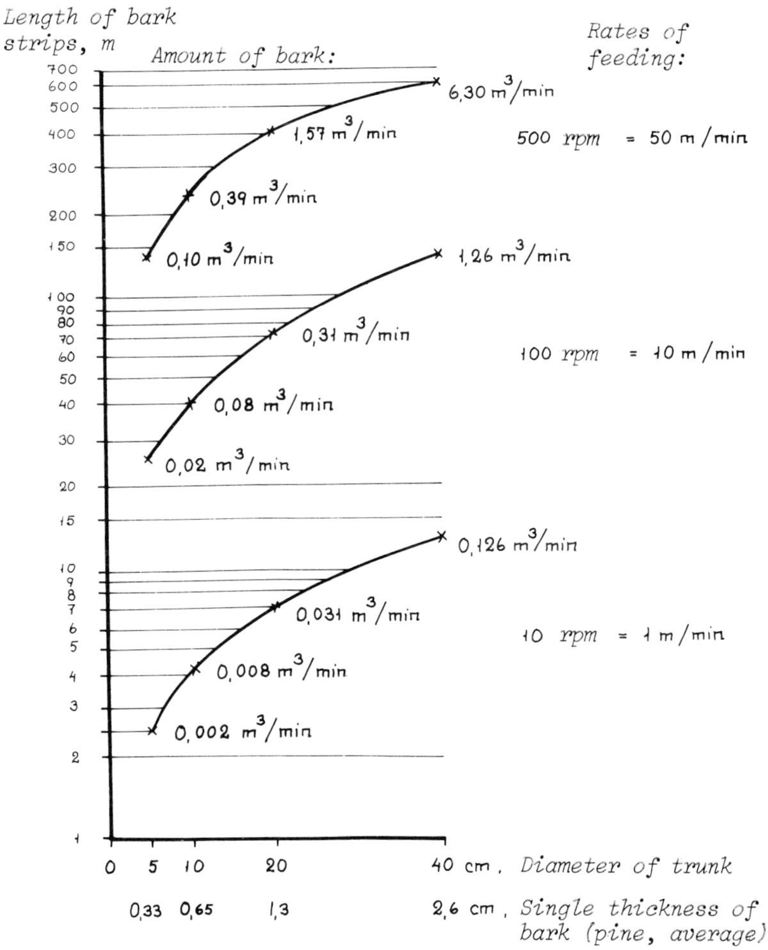

FIGURE 123. Debarking capacity at various diameter of the trunks and varying rates of feeding (Staaf, 1972).

- Since output is influenced strongly by various diameter, the manufacturers have chosen to market series of machines where each unit has a given rate of feeding for each range of diameter.
- However, it would be valuable from the point of small timber economics to achieve a higher output per unit of time by a regulation of the rate of feeding i.e. an automatically variable rpm or width of the debarking tools.
- Economics of debarking small timber is particularly critical because of a relatively large bark surface, see Figure 124 (Sundberg, 1957).

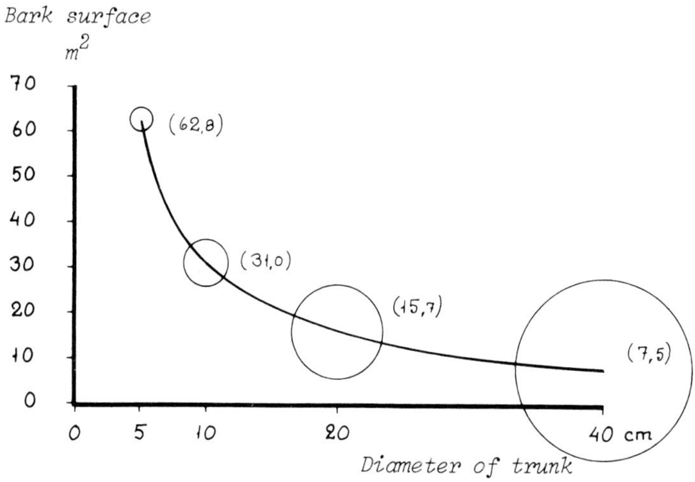

FIGURE 124. Relationship between bark surface and diameter of trunk

Hydraulic debarking in the forest

Hydraulic debarking in the forest does not occur to any practical extent. It would require too large machine units. However, the principle is applied at industries (Harris, 1965).

Chemical debarking in the forest

Chemical debarking of standing trees is done by injection of chemical substances into the trunk. The tree is killed and the bark is loosened or removed without great resistance by regular debarking procedures. The method is of no practical consequence which has been demonstrated by several experiments i.a. in Denmark (Moltesen, 1965).

Debarking integrated with other harvesting operations

Mechanized debarking in the forests is fraught with crucial problems concerning organization and transport technology which often require special solutions on the basis of studies of methods and analyses of costs. This applies in particular to the case when debarking is done as a separate operation in scattered locations.

Factors influencing the result of debarking

To attain the maximum production, it is necessary first to arrive at an optimum combination of a large number of factors that vary from one debarking location to the other. An optimization of the factors is rarely possible in practice (Staaf, 1965c).

The following factors that influence the result of debarking may be mentioned:

Organization:	Location of debarking
	Labour
Amount of timber:	Size of storage
	Distance between various places of storage
	Length of piles
	Height of piles
	Distance between piles
Timber:	Average length
	Average volume per piece
	Species
	Crookedness
	Temperature of bark
Machine equipment:	Rate of feeding
	Operational dependability
	Cost of operation

Reasons for integrated debarking

If debarking can be included in a wholly mechanized harvesting system where it is integrated with several other partial operations, some time consuming organizational work can be avoided. To some extent this organizational work is done once and for all in conjunction with the construction and the design of the work procedures.

The matter of integrating debarking with other processing of timber can be judged from other points of view as well. It may be a feasible solution to integrate debarking at landings into a processing establishment for reasons of transport technology e.g. in areas of river drives.

The economic advantage of integrated debarking increases particularly in cases when the trunks are processed in an establishment with a relatively high production.

Integration of debarking with other operations in the forest or at landings is increasingly compared with the relatively inexpensive industrial debarking which in turn is more or less integrated with the manufacturing process.

Distance of transport is the crucial factor largely influencing the choice between debarking in the forest and debarking at the industry. The closer to the stump debarking can be done, the lower will be the cost of transporting the wood, the transport of bark being eliminated. A reduced cost of transport of dry, low weight wood can be expected. However, a lower cost of debarking is experienced at higher debarking capacity, the amount of timber available for debarking increasing closer to industrial sites. Figure 125 shows the relationship between cost and distance of transport.

When the difference in cost between debarking in the forest and debarking at industry increases, the point of interception between the curves for transport cost of debarked and rough timber moves to the right, i.e. rough timber can be transported over longer distances to the industry. When the difference in cost between transport of debarked and rough timber is reduced, which often occurs at increased transport capacity, e.g. larger trucks, the point of interception is also moved to the right, and rough timber can again be transported over a longer distance to the industry. These relationships are some of the main reasons behind the increasing extent of industrial debarking.

Bark may have a value as fuel at certain industries. However, increased industrial debarking could lead to a greater risk of insect damages in our forests, resulting in losses of growth.

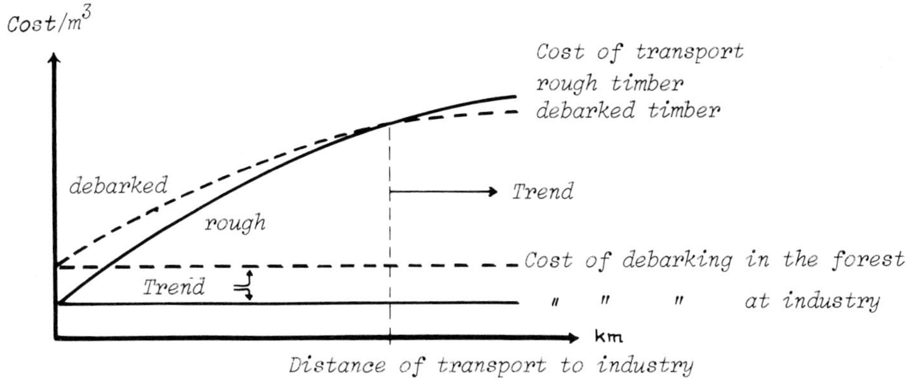

FIGURE 125. Relationship between cost of transport and distance. Rough and debarked timber.

Types of debarkers

Debarking can be organized for and done by means of stationary, semi-mobile and mobile units.

Stationary debarkers

Stationary debarkers are machines or establishments of a permanent nature, either connected directly to industries or installed at major central establishments in combination with other handling or processing of timber e.g. sorting and chipping.

At virtually every sawmill in Sweden there is a stationary unit for debarking of timber, i.e. with possibilities for chipping of the slabs into material desirable in the pulp industry.

Semi-mobile debarkers

Semi-mobile debarking machines can be moved, albeit at a relatively high cost, to a few work locations per season or per year. This category includes machines for debarking at major landings.

Mobile debarkers

Mobile or portable debarkers are often mounted on a tractor, a truck or a trailer. The units, therefore, are easy to move and they can be transferred at reasonable cost between several work locations in a season or year. This category primarily includes machines for debarking along truck roads and on small landings.

Most common forms of organization at debarking by means of small units

Debarker mounted on tractor

The tractor is also a power source for the debarker. Infeed and outfeed are manual. The unit normally requires a crew of three persons. Production per shift is approximately 130 m^3 of piled wood.

Debarker mounted on tractor-trailer

Infeed and outfeed of timber is done by a hydraulic crane mounted on the tractor or on the trailer.

While rough timber can be kept on both sides of the unit, debarked timber is unloaded on one side only (outfeed side), debarking being done across the machine. Three workers are a normal crew. Production per shift is approximately 170 m^3 of piled wood.

Mobile debarker

This debarker is mounted on a flatbed or on a specially built trailer. Timber, which is handled by means of a hydraulic crane, can be kept on both sides of the machine. Three-men crew. Production per shift is approximately 200 m^3 of piled timber (Figure 126).

A big mobile debarker with high capacity and of interest from a technological point of view was built in 1966. The unit was developed from a previous debarking machine and it is now mounted on a truck chassi with a load capacity of 10-12 tonnes. In addition to the debarking unit it consists of two cranes, a hydraulically maneuverable infeed deck, bark blower, outfeed mechanism and outfeed bin for debarked trunks, and a centrally placed operator's cabin. The unit is operated by two men in such a way that one man in the operator's cabin maneuvers both cranes, one for rough timber and one for debarked timber. The rear crane lifts rough timber from a pile onto the infeed deck where the second operator is responsible for work with the infeed conveyor (Cf. Muszynski, 1976).

The infeed mechanism can control the rate of feed continuously by means of hydraulic equipment, utilizing fully the capacity of the debarker under various weather conditions and for varying sizes of timber. Too big timber or other timber unsuitable for debarking may be fed into a bin from which the crane is unloading the timber on suitable occasions.

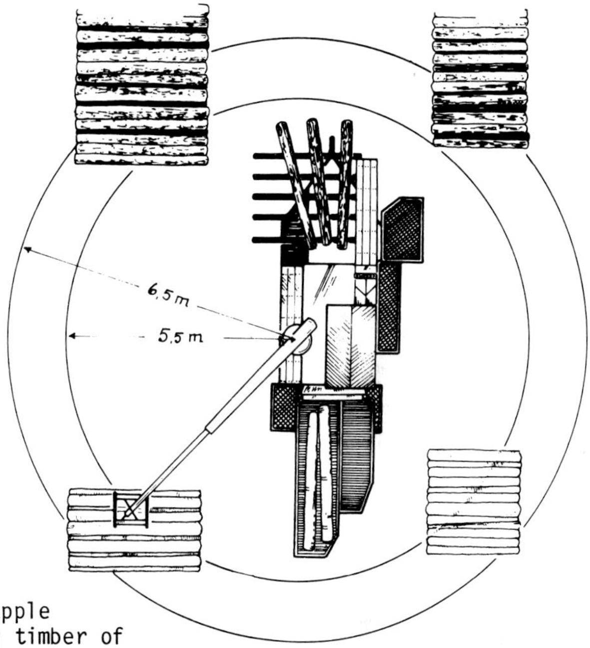

FIGURE 126. Mobile debarker equipped with grapple loader processing timber of two and three metre length.

Debarking of pulpwood in troughs

The method involves the placing of about 15 m^3 (stacked) rough pulpwood in a trough, the bottom of which is comprised of revolving screws. These screws force the pulp wood to move around the trough. Through friction, partly against the sides of the trough and partly against each other, the bark is removed from the pulpwood. To a large extent, debarking results depend upon the time in which the pulp wood lies in the trough.

At present, trough-debarking as a method cannot be regarded as being completely developed. Among other things, a number of ergonomic factors can be improved e.g. noise level. However, it should be emphasized that the method is already, from an ergonomic point of view, far superior to the older motor technique (Dehlén et al, 1982).

FIGURE 127. Debarking of pulp wood in a trough.

Development trends in debarking

Increased mechanization in a chain of production leads to higher productivity. Simultaneously, the capital requirements for procurement of machines also increase. Thus, the cost structure changes, i.e the proportion of costs for manual work declining in relation to the costs of capital and operation.

Relationship between the cost of manual work and degree of mechanization

Cost of labour at manual debarking appears to be 99 percent of the total cost while the cost of equipment is maximum 1 percent.

At debarking by means of small machines the cost of labour is estimated to be approximately 75 percent while the machine costs amount to 25 percent of the total cost of debarking.

If a bigger machine with hydraulic crane and a service crew of four is used for debarking, the proportion of labour costs declines further to about 50 percent, while the remaining 50 percent are machine costs.

When debarking is done by means of a big mobile machine, with a 3-men crew and an output of 25 m^3 of piled wood per hour of efficient work, the cost of labour may decline further to 25 percent while the machine costs increase to 75 percent. The cost of personnel at debarking by means of thumbler in a pulpmill is estimated to be 5 percent of the total cost of debarking depending on type of thumbler. At hydraulic debarking in establishments which are operated by one man and producing up to 1700 m^3 per

8-hour shift, the machine cost exceeds 95 percent for depreciation, interest, energy and maintenance.

However, a change in the cost structure does not indicate what happens to the total cost. In a region with very low wages, high capital costs, expensive machines and high costs of energy, e.g. in a developing country, manual debarking is cheaper than mechanized debarking.

In two countries having about equal wage level but different costs of energy and transport tariffs, e.g. a comparison between West Germany and Sweden, it is likely that debarking by means of a big machine in the forest is advantageous in one country while debarking at industry is advantageous in the other country (Steinlin, 1969).

Cost of labour climbs faster than machine costs

It has been shown on several occasions that the cost of labour is rising faster than the costs of machines and energy. For this reason the forest operations must shift into a higher degree of mechanization.

In the beginning of the 1950's manual debarking was no doubt the cheapest method of debarking. In the middle of the 1950's it appeared that small debarking machines were interesting from an economic point of view. A machine mounted on a tractor experienced its break-through as a mobile unit in the forests in the end of the 50's. It was replaced with high capacity large mobile units and semi-mobile establishments with hydraulic cranes for infeed and outfeed.

Today industrial debarking, predominantly done in machines with thumblers, is the best economic alternative in spite of the transport cost of bark and despite relatively long transport distances.

Investments required

The acquisition of a debarking spud is paid off if the annual production is 10 m^3. The purchase of a small debarking machine may be justified if the annual production is estimated to be 2000 m^3 - 3000 m^3 while the purchase of a debarking machine with rings may require an annual production of 25 000 m^3 - 30 000 m^3 to be justified.

To be viable an establishment with debarking thumblers may require a production of approximately 200 000 m^3 per year.

In many cases it may be enough to produce 25 m^3 per hour, or 70 000 m^3

per year if debarking in thumblers is to be a viable operation. If sufficient quantity cannot be produced, establishment will be operating with too high costs. This situation manifests itself when the establishment appears to be less useful than an establishment with a slightly higher technical development that can be utilized fully. If the amount of work is insufficient, activity is forced to assume a lower level of mechanization with higher costs.

Trends in concentration

The situations described above have brought about a trend towards concentration in all business activities. In this context the forest operations are in a relatively disadvantagous situation, having limited possibilities to increase production.

For natural reasons the forest operators cannot harvest an unlimited amount of timber within a given area. If useful machines are desired in forest operations, timber must be either collected from a large area in order to reduce the cost of processing bought at a higher cost of transport, or mobile machines must be put into operation over long distances. Still, time is spent on expensive moving which reduces the utilization of machine capacity. This situation is causing higher processing costs, in this case the cost of debarking, per unit of volume. The disadvantage associated with the limited yield per unit of area is further aggravated by the occurence of many different species, assortments for various purposes, or various methods of manufacture. It may also be exacerbated by a relatively low proportion of forest land severely fragmented into farm land, urban areas, lakes and waste lands, particularly in forest areas within farming districts.

Regions with continuous areas of forest land, e.g. up to 80 percent of the total area and with a few species giving one or two assortments such as parts of Canada and Russia, have a much greater potential for viable tree harvesting operations than have the middle European countries with their higher yield of timber per hectare.

Another obstacle to increased mechanization is the occurrence of many small forest ownerships and the fragmentation of properties. Many owners of small holdings are not in a situation where they can use machines and stationary establishments to the full extent. The small ownerships, therefore, are forced to accept a lower level of mechanization in their operations and higher costs of tree harvesting. The small owners are also liable to experience more strongly the rising costs of personnel. To alleviate this disadvantage they increasingly collaborate within the frameworks of forest management areas and similar organizations.

Bunching

Bunching of timber is a transport operation carried out in a harvest area for the purpose of collecting timber into concentrations e.g. by directed felling or by manual bunching. Thus, bunching is usually closely associated with the other partial operations of harvesting and it is, therefore, mentioned in the description of these operations.

Two forms of bunching can be distinguished viz. manual bunching and bunching by means of horses or tractors, besides directed felling.

Manual bunching

Manual bunching is a collection of timber by hand.

Bunching by means of horses or tractors

Bunching can also be carried out by means of horses or tractors. The term 'scooting' is used in certain areas for bunching of timber into piles.

When a grain field is harvested, the farmer collects the straws into e.g. sheaves, and at threshing he collects the grains in sacks. Collection of timber in the forests is principally the same thing but on a different scale.

Purpose of bunching

The purposes of bunching are primarily to achieve a rational processing, to facilitate a subsequent transport operation, and to protect and store timber properly.

A more efficient processing can be achieved if e.g. the trees are felled in such a way that they can be processed further in bunches at delimbing and bucking operations. Even when the timber is delimbed only, it may be rational to collect the trunks for bucking in bunches.

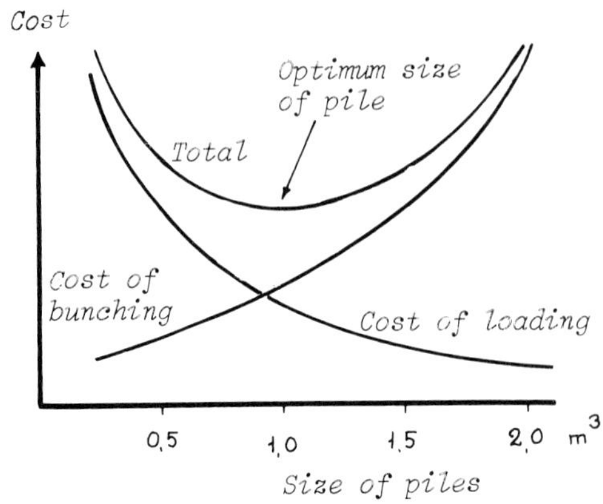

FIGURE 128. Optimum bunching of timber

Collection of small trees, trunks or parts of trunks into bunches is primarily done in order to obtain processing units of optimum sizes for the subsequent operations. Purpose is the same with respect to bunching done to rationalize transport work.

Directed felling provides a certain degree of transport in a desired direction. It also facilitates largely the bunching process which is done i.a. for the purpose of making suitable units of handling or bunches for loading at further transport.

An optimum size of bunches is desirable when timber is collected for transport. This size depends on the cost of bunching which rises with increasing size of the bunch because of the longer transport distances necessary for the collection of trees or trunks. The optimum size also depends on the cost of loading which declines at increasing extent of bunching, concentration of timber, or size of the bunches. The optimum extent of bunching is shown in principle in Figure 128.

Bunching may also be done to protect the timber, for instance by storing the timber in various types of piles (triangles, crosses, crates or in piles with stickers) in order to obtain a quick drying. Storage of timber can be arranged in the stand or along the roads. This form of timber storage requires debarking in order to prevent damage from insect infestation or storage decay. However, storage of timber in the forest has declined in recent years.

Other purposes of bunching may be to achieve a concentration of timber quantities for e.g. central processing, to obtain a buffer storage between various links in the transport chain, or to accumulate timber before delivery.

Work techniques and equipment

Manual bunching

Work techniques and equipment used at bunching of timber vary between different forms of collection. Manual bunching is combined with directed felling and the pattern of felling depends on the distance of bunching.

Manual bunching over relatively long distances can often be facilitated if the trunks are hauled on top of other trunks to e.g. a strip road. Equipment used at manual bunching usually consist of lifting tongs or lifting hooks.

Bunching by means of winch

Bunching can be done by some type of haul, previously by horse, more recently by means of winches and/or tractors. This operation is discussed further in the section on transports. Trees or trunks can be hauled for processing at the road side by means of a separate winch. Since the cost of bunching by means of winch is lower than the cost of manual bunching at quantities above a certain minimum, distance of winching can be extended correspondingly. This form of bunching provides for a higher concentration of timber at the strip roads which will facilitate loading and transport.

Bunching by means of crane

Bunching can be done by means of a tractor equipped with felling mechanism on a hydraulic knuckle boom or a straight telescoping boom, for felling of one or several trees simultaneously. The pattern of operation varies between different forms of mechanized felling.

Bunching by means of processing machines

Other machines can be used for bunching of more or less processed timber.Thus a delimber can be used for bunching for instance along strip roads to facilitate bucking or further transport.

Machines for delimbing and bucking often carry out bunching more or less automatically in cradle like attachements which are unloaded along strip roads and on landings, or loaded directly onto flatbeds of trucks.

Various forms of bunching

Equipment and work methods applied for concentration of timber combine into various forms of bunching. The forms can be put into two main groups: manual bunching and mechanized bunching.

Manual forms of bunching

Rush felling. Bunching of bucked timber is carried out by a separate worker when timber is hauled by horse. This is an old form of bunching.

Planned felling. In contrast to rush felling, planned (directed) felling provides for essentially facilitated bunching. This is obvious if trees are felled towards, instead of away from, the place of processing or a strip road.

In stands the timber may be put into bunches containing 3-6 trees or trunks or into piles of various types e.g. crosses, triangles or crates. This form of bunching is no longer common.

Timber at strip roads can be bunched into strings, piles, bunches, crosses or crates. The choice depends on i.a. the amount and length of timber. Usually, timber is put into piles or bunches the sizes of which are determined by the lifting capacity of the loading equipment used for forwarding.

Bunching of bucked timber (assortments)

Bunched timber may be transported to landings for unloading into storage space or directly on the flatbed of a truck. The choice of unloading place is influenced by i.a. transport distance to landing, type of loading equipment and transport means available.

At long transport distances separate carriers are used for the bunched timber. It is only when distances are very short (approximately 100 m) that processing machines can be used for transports of the bunched timber.

Bunching of tree length trunks

Bunching in combination with delimbing

After delimbing of standing timber, the trees are felled and put into bunches or piles. After delimbing of felled trees, the trunks are placed into bunches or piles along strip roads or in swaths.

Bunching in combination with bucking

This method may be applied e.g. by means of separate timber cradles, often for assortments, mounted on the processing machines along the strip roads. Bunching can also be done by means of front end loaders at bucking of bunches or at bucking of entire piles of trunks at landings.

Bunching of trees

Bunching in combination with transport

Bunching in combination with transport of trees can be done immediately after felling by means of a feller - buncher or a feller - skidder. This form of bunching can also be carried out in difficult terrain by means of winches and cable lines e.g. cable cranes and high slack lines.

Bunching in combination with processing of trees

Timber is bucked after delimbing and (bunched) sorted into pockets along a ramp or into assortment cradles.

Bunching in combination with processing of bunches

Bunching is done for instance when timber is bucked and delimbed in bunches.

Bunching integrated with other harvesting operations

The previous section on various forms of bunching shows how this work can be carried out and integrated in many different ways. The optimum form of bunching gives the lowest total cost if it is combined with other partial operations in the harvesting system. Analyses of various man-machine methods, performance and costs are necessary. These matters are discussed further in a subsequent section.

Trends in bunching

Bunching of timber is physiologically very heavy work that should be done by means of machines.

One trend, therefore, is to replace manual bunching with mechanized operation to the greatest extent possible.

Another trend is a development toward increased handling of timber in bunches, particularly for small trees and trunks. This development is promoted by an increased interest in the bunching process.

In a system consisting of felling, forwarding to terminal and centralized bucking of the trees into sections, bunching and delimbing of the sections of several trees at a time can be done in special machines.

Bunching and delimbing the trunk sections, the machines are designed to reduce the total cost of tree harvesting and, in combination with chipping, to facilitate the recovery of limbs and tops in large amounts for fuel purposes.

The machines can bunch several trees simultaneously and it is considered to be a component of the system for handling of trunk sections or tree parts. The system is also very suitable for processing of timber at sawmills and pulpmills.

Description of a machine for bunching-delimbing of trunk sections - 1983 model

Bunched trunk sections of 6-7 m length are delimbed in a trough that can hold 2-4 m^3 of limby timber.

In the bottom of the trough there are holes that can be opened hydraulically in order to discharge the short, broken pieces of timber accumulating between the chain conveyors. In the lower part of the trough there are three punching rollers and one knife roller where delimbing is done. The chain conveyors (mentioned above) are designed for turning over and removal of the trunk sections in combination with welded-on 250 mm long arms for lifting of the timber. The chains are operated by means of hydraulically powered drives.

Discharge of waste, limbs, nobs, bark and small broken pieces of timber is done on the short side of the machine by means of roller feed downward to the bottom of the trough via guiding plates onto a conveyor belt. Timber is bunched for delimbing from a separate infeed bin which can be maneuvered by hydraulic cylinders. During transfers from one location to the other the bin can be folded into the trough since it has approximately equal timber holding capacity.

The machine is operated from a cabin by means of mechanisms maneuvered by hands and feet.

Movements of the machines over short distances can be operated directly from the cabin by means of manually activated valves controlling speed and steering.

The machine is equipped with a crane having a loading capability of 135 tonne-metres and a reach of 9.5 m.

Movements on the terminal (or the industrial site) are made by means of hydraulically powered wheels. For long transfers on the roads the wheels are removed and the machine is placed on a trailer for conventional transport by truck.

Performance of the machine depends on several factors such as characteristics of the timber, method of piling, operational manners and auxiliary equipment.

FIGURE 129. The first version of the buncher-delimber. More recent models being manufactured and modified are improved in some details. (Photo Kurt Svensson).

Over 100 m^3 solid wood can be processed per hour of efficient work. In combination with debarking the output will be strongly reduced to approximately 50 m^3, for timber of spruce and deciduous trees even lower. Higher performances are expected after additional improvements of the machine, which has been studied by the Faculty of Forestry in Sweden in 1983.

Increased handling of timber in bunches should lead to a reduced sensitivity to diameter in the current harvesting systems which might become of particular importance for thinning operations in young stands.

Chipping

The processing of chips in the forests for further transport to industries has not yet become common practice. The method would be of great interest if the cellulose industries were prepared to accept chips containing needles, twigs, bark and parts of roots.

If chips are to be used for energy purposes, a new system of thinning, chipping of entire trees in the stands, could be introduced. In addition, the residues from cleaning and from harvesting operations in mature stands could be utilized.

Needles, bark and cones

No comprehensive laboratory tests with pulping of forest chips containing needles and bark have yet been carried out. However, it is known that approximately 55 percent of the total amount of cellulose in a tree is contained in the normally utilized parts of the tree, 3 percent in the top, 17 percent in the limbs and needles, and 18 percent in the roots.

It is also known that the trunk has the longest fibers and the highest content of cellulose, that the tops have a slightly lower content of cellulose but a good fiber structure, that the big roots are good raw material and that the limbs have a slighly lower amount of fibers. The main problems in the pulping process are caused by the needles, cones and bark. However, there are technical possibilities to screen the chips in order to achieve the quality required for a good pulp.

What is chips?

Chips consist of woody material that has been mechanically fractioned into pieces of a size suitable for e.g. pulp, fiber boards or fuel. Cellulose chips, fiber chips and fuel chips, therefore, can be identified.

Chipping - fuelwood

As a result of the oil "crisis" in 1973 and the ensuing sharp increases in the prices of petroleum products, it has become clear that wood is a potentially valuable substitute for the non-renewable and expensive oil (Wiksten, 1977). This development is of particular importance in forested countries which depend on imports of petroleum products for their domestic needs. Wood, bark and foliage (biomass) from low quality trees or parts of trees now have a value corresponding to the energy equivalent amount of oil, coal and natural gas. It has become economically advantageous to utilize small trees from cleaning operations as well as limbs and tops from all tree harvesting operations and to produce and utilize fuelwood from energy forests of fast growing deciduous species. Stumps and peat have also become valuable energy alternatives.

This sudden renaissance of the usefulness of wood has brought about a rapid development of tools and equipment for chipping of trees and parts of trees from cleaning, thinning and final harvest operations in the forests and for the centralized processing of this material at landings or industries.

In addition to being an important stand improvement measure, cleaning now, particularly in a late state, supplies material of great interest as a source of energy for heating. The main problem in the development of machines and methods is to design equipment that is both efficient and easy to use in the young stands without causing damage to the remaining trees and the sites.

Throughout the world from the harsh northern and alpine regions to the steaming, hot jungle with remnant bush vegetation in the tropical forests, there are numerous, neglected stands which now can be gainfully treated with improvement measures while being an important source of domestic and renewable energy (Cf McMillin, 1978).

Types of chipping machines

Chippers can be classified into two main types: chippers equipped with cutting knives or with slashing knives. The latter type is normally not portable.

Chippers with cutting knives can further be divided into the following main types:

- Disc chippers
- Drum chippers
- V-form chippers

Disc chippers are best suited for the production of high quality industrial chips.

Chips obtained by drum chippers are generally of a lower quality, but some drum chippers can produce good quality chips. Drum chippers are very suitable as portable machines because of their light weight and wide feeding spout. A great many of the so-called small-sized wood chippers are of the drum type. They are better suited to chip undelimbed timber than disc chippers (cf. Anon. Joint Committee, Log 161, 1965).

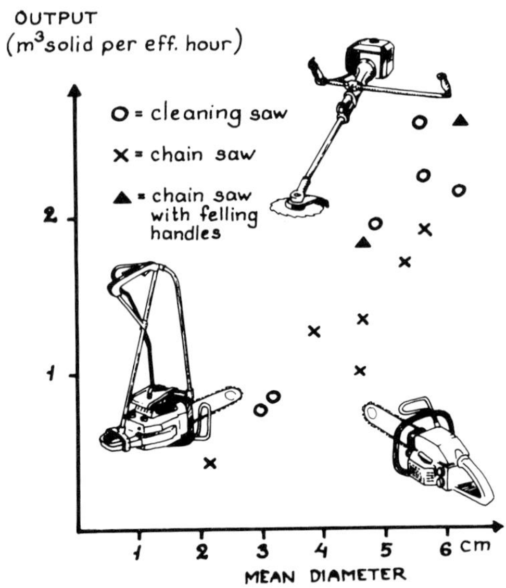

FIGURE 130. Output of chips at cleaning depends on tools used and on the size of trees removed.

Chipping with portable chippers

Before the stand is cleaned motor-manually by means of chain saws or circular saws, a net of strip roads has been laid out. Cleaning can also be done by means of chain saws equipped with felling handles.

Felling. Output at cleaning (excl. piling) depends on the tools used and on the size of trees removed (Figure 130).

Figure 130 shows that the cleaning saw is most efficient. However, if felling is combined with piling of the trees, a chain saw with felling handles is superior since it will facilitate directed felling. The felling handles are auxiliary equipment providing a clearly improved working position and reduced hazards.

Chipping in the forest

The trees are felled with the top or butt ends pointed toward the strip road.

Chipping of trees by means of small portable chippers can be done while the vehicles are moving in the stands. An example is given on the use of a portable chipper in a stand designated for cleaning (Filipsson, 1983).

The chipping vehicle consists of a tractor as base machine, a chipper mounted in front and a high tipping hopper with a capacity of approximately 5 m^3 behind the operator's cabin (Figure 131).

Output of chips may amount to 5.5 m^3 - 6.0 m^3 (bulk volume) per hour for trees with a dbh of 6 cm - 7 cm and for a transport distance of 200 m. At an annual production of 6 000 m^3 of chips, the costs of operating the machine in 1983 was calculated to be $3.00 (U.S.) per m^3 of chips.

To produce this output the following conditions are necessary:

- good planning
- efficient maintenance of tools and equipment
- quick selection of trees
- directed felling

The advantages of this method of chipping in young stands are:

- chips are produced from the cleaned stands at a relatively low cost
- slight damages to the remaining stand
- reduced insect hazards
- clean chips, free of soil and rocks
- relatively dependable method of chipping
- chipping unit can be moved easily between various positions

FIGURE 131. Vehicle equipped with chipper and chip hopper.

Distance between the strip roads is 20 m. The short chipper unit can be easily moved in the stands passing through openings without damaging the marginal trees along the strip roads.

Trees from cleaning are bunched manually and fed by means of hydraulically powered rollers into a chipper while the operator collects more cleaning residues.

The chipper is operated at 1 000 rpm. The chips are ejected through a duct to a hopper mounted on the rear of the tractor. When the hopper is full, the chips are transported to a truck road and tipped into a large container for further transport by truck.

Description of chipper:	Power requirements:	20 kW at 540 rpm
	Cutting disc, diameter:	800 mm
	Weight:	136 kg
	No. of knives:	2
	No. of counter edges:	2
	Infeed opening:	200 mm x 200 mm

The chipper can be positioned straight forward or slightly to the right or to the left in front of the base machine, thus influencing the work process for cleaning-chipping and the planning of strip roads. To a large degree performance of chipping depends on the number of trees chipped at each position.

FIGURE 132. Unloading of hopper into a chip container.

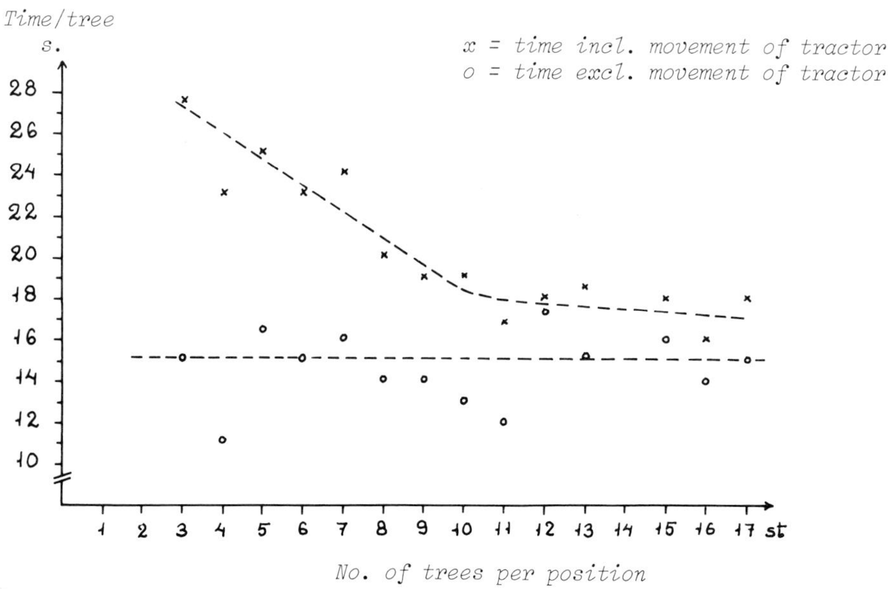

FIGURE 133. Time of chipping, s/tree, depending on no. of trees/position.

Output also depends on the speed at which the tractor can travel in terrain and on the distance to containers at the truck road.

Travel speed in terrain is usually approximately 75 m/min, with or without load. Distance to truck road may be 500 m or longer.

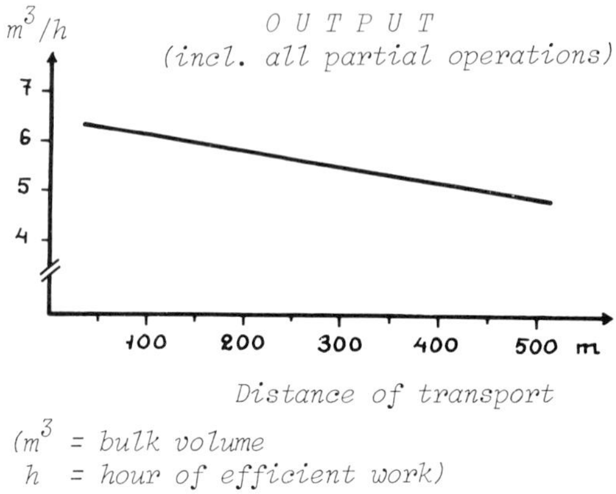

FIGURE 134. Influence of transport distance on the output of chips.

Chipping of residues from thinning operations (tops and limbs for fuel)

Tractor mounted chipper

This chipper has two hydraulically powered infeed rollers. The chipper is capable of processing limbs, tops, limby timber and whole trees from thinning operations.

Technical data:

Infeed opening:	250 mm x 250 mm
Size of chips produced:	5 mm - 12 mm
Output capacity:	10 m^3 - 15 m^3/h
Power requirements:	30 hp - 100 hp
Weight:	680 kg

FIGURE 135. Chipper with two hydraulic infeed rollers.

This chipper is equipped with an infeed mechanism which can be used whether timber is frozen, dry or limby. It has a trailer coupling gear facilitating infeed horizontally and obliquely. The chips can be ejected directly into the trailer.

The chipper is available in three designs: with mechanical or hydraulic infeed mechanism and without infeed mechanism.

Chipping of energy forests and tree harvesting residues requires efficient equipment

The basic concept is an operation of mobile chippers close to the growing sites. This mode of operation will yield chips with a minimum admixture of impurities (gravel, sand etc). Clean and uniform chips burn efficiently with a minimum of disturbances. A mobile system of chipping may consist of the following equipment:

- chipper
- trailer with chip container
- containers for truck transport

FIGURE 136. Chipper with hopper.

FIGURE 137. Chipper with trailer.

Coordination of the various partial operations

Discussing the various partial operations, we have realized how they are interdependent of each other and that they are all influenced by the operational procedures chosen. From felling at the tree sites through the following partial operations we have noticed that most of them contain elements of terrain transport. We will now review briefly coordination and the modes of production applied for the achievement of the specific objectives.

Coordination

Planning of tree harvesting in detail cannot be summarized easily in a few words, work being strongly dependent on suitable methods, technical means, extent of work, terrain conditions etc. Planning in detail, therefore, has also been discussed in the context of various types of work, operations and methods. The main purpose of work planning is the achievement of systems that require the least possible input of labour with minimum risk to health and life and with maximum possible performance.

Rational coordination

A wide choice of technical means is now available for tree harvesting. It is an important matter how these means should be coordinated and utilized rationally in a varying environment of work.

It was natural that the technical means first became useful for the handling of heavy trees and units of timber. Modern technology has also proved advantageous in areas with a large volume of timber per hectare e.g. clearcutting of old stands. To some extent whole trees or trunks are now being transported to landings or to a central place with more or less industrial processing of timber.

It is also natural that work input per unit of volume is higher for small trees than for large trees. Figure 138 shows this relationship for manual felling and processing. Input of time per unit of volume is over 2.5 times higher for a 5-cm (Dbh) tree than for a 10-cm tree. The same relationship applies to operations of single-tree machines in which output of harvest increases in proportion to the increase in diameter squared, at constant rate of feed. (Staaf, 1965b).

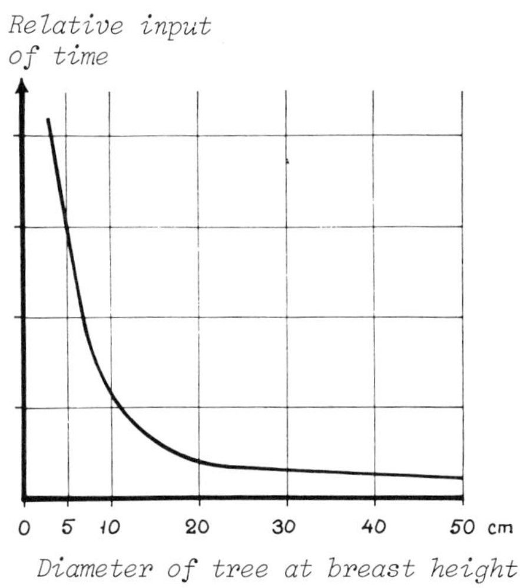

FIGURE 138. Principal relationship between input of time per unit of volume and tree size at single-tree felling and processing.

Objective of production

Production of timber should be aimed at achieving the best economic result by means of the least possible input of resources at the lowest cost. This objective is achieved when the relationship between the cost and quantity of production gives the lowest cost per unit of product.

The relationship can be expressed in the following simple way. (Staaf, 1965b).

Formula 1.

$$\frac{\text{Total annual cost}}{\text{Total annual production}} = \text{cost per unit of product}$$

Formula 2

$$\frac{\text{Cost of capital + cost of operation}}{\text{Degree of utilization x production}} = \text{cost per unit of product}$$

An equally low cost per unit of product can be achieved theoretically by means of several different altenatives e.g. production by means of expensive multi-process machines giving a high output, or production by means of relatively inexpensive single-process machines giving a lower output.

However, some fundamental factor usually determines the choice of production alternative e.g. the amount of timber that is to be processed. Industrial processing may be advantageous when timber is collected at a central place (depot) while more simple operations are feasible when timber is scattered in small amounts.

It may be questioned what the possibilities are to influence conscientiously the factors included in Formula 2. The question is quite appropriate in the case of e.g. timber of relatively small sizes from thinning where volume of timber removed per hectare is low.

In view of the current trends concerning cost of labour, the matter of thinning has become very critical, the input of labour increasing sharply when the amount of timber removed per unit of area and the diameter are small.

To give an answer to the question concerning the possibilities to adjust the factors in Formula 2, it is necessary first to analyze the modes of production that may be considered.

Various modes of production

Processing and transport of timber may be either continuous or discontinuous. The partial processes included in production can be coordinated in sequence, parallel to each other or in some other combination. (Staaf, 1965a).

Systems of various modes of production

The following systems of various modes of production can be designed:

1. Continuous production is a mode of production without interruptions for other work phases such as infeed, transfers, unloading etc.
   - 1.1 Partial operations coupled in sequence. The partial operations are carried out in sequence without interruptions.
   - 1.2 Partial operations coupled parallel to each other. Several partial operations are carried out simultaneously such as felling, delimbing, bucking and bunching.
2. Discontinuous production is a mode of production with frequent interruptions for other work phases such as infeed, transfers, unloading etc.
   - 2.1 Partial operations coupled in sequence
   - 2.2 Partial operations coupled parallel to each other.

A continuous production in which the various partial operations are carried out parallel to each other gives a high utilization of the capital investment (machinery) and a high performance, normally with a low input of labour.

A discontinuous production where the partial operations are carried out in sequence, gives a low utilization of the capital investment (machinery) and low performance, normally with a high input of labour as in conventional, manual work (Figure 139).

Continuous systems with parallel coupling

Production modes now being developed in forest operations appear to evolve into continuous man-machine systems with parallel coupling.

For relatively small trees (approx. 10-15 cm dbh) from thinning in young stands, various partial operations can be coordinated in a harvesting machine for continuous operation.

In contrast, harvesting in old stands with big (approx. 30-50 cm dbh) trees and, hence, heavy units of handling, is for reasons of volume and weight difficult to carry out by parallel coupling of partial operations.

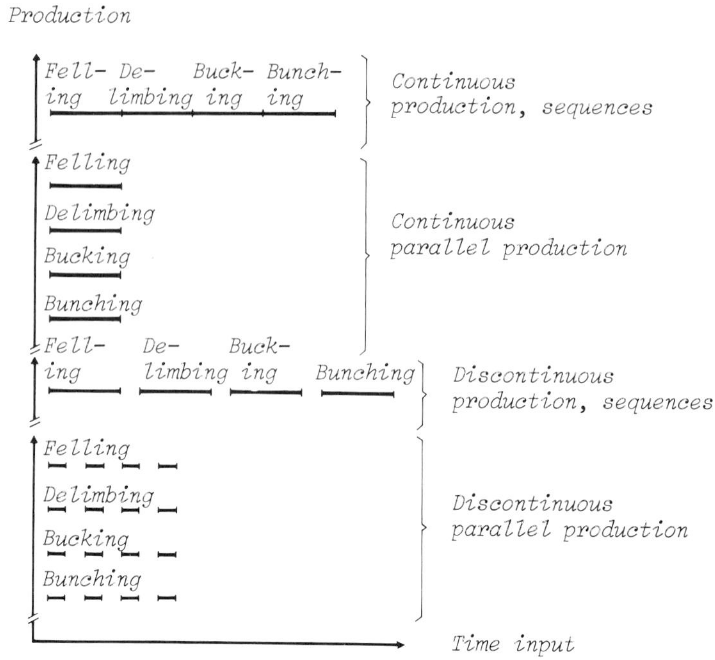

FIGURE 139. The partial operations can generally be coordinated in various modes of production. (Staaf, 1965b).

Costs of capital and operation

The cost of a harvesting machine for continuous production with parallel coupling of the partial operations may be reduced due to concentration of equipment. This concentration can be achieved by the use of common components such as machine frame, primary source of power, oil pump, operator's cabin, transport components, maneuvering mechanism, protective equipment etc. A concentration of equipment should also serve to reduce the costs of operation.

The cost reducing concentration of the machines is difficult to apply when production is based on several different machines.

The need for labour such as operators, loaders etc. is less for multi-process machines with parallel coupling of the partial operations than in systems with several single-process machines.

Modern technology certainly provides great possibilities to influence the cost factors in Formula 2.

Utilization of equipment

Utilization of equipment or the time of efficient operation (h/year) in percent of total time available varies among the modes of production. A relatively high utilization or a short time of production is achieved where machines for continuous systems with parallel coupling are used (Figure 139).

A relatively low utilization or a long time of production is achieved in modes of production using machines for discontinuous operation coupled in sequence.

Two- or three-shift operations provide a high annual utilization or a long time of production in relation to the costs of capital and operations.

To achieve a high utilization of equipment at felling and processing in the forest, it is necessary to move the machine continuously towards the standing trees. In contrast timber is moved toward the machines at centralized operation.

Continuous processing along a strip road, which is technically possible today, will make productivity less dependent on the quantity of timber felled per unit of area. Principally, the slow movement of a harvesting machine along a strip road is only the reverse of harvesting with a stationary machine towards which the trees are moving. This operation provides a certain degree of independence of the quantity of timber felled per unit of area - a relationship worth noticing.

Production

High production ("hot logging") is achieved when several partial operations are run with parallel coupling. Production and processing at felling varies largely with the size of the trees.

However, influence of tree size on production could be reduced or eliminated by means of processing machines if e.g. rate of infeed is increased automatically for trees of small diameter. This arrangement would mean that infeed and other engines could work with a more stable performance (constant load). A constant level of load should lead to a better production result than that experienced with strongly varying requirements for engine output, characterizing a number of machines today. This is a detail of production pertaining to the matter of smallwood harvesting. There are now technical solutions to this problem, e.g. some processors have an automatic infeed rate regulated for the process of delimbing by the size of the tree.

Thus, all factors in Formula 2 can be influenced and adjusted in order to achieve production at a lower cost per unit of timber.

If a partial operation can be integrated into a man-machine system where several other partial operations are involved, a considerable part of the corresponding time consuming organization can be avoided. Actually, this organizational work was already carried out in several respects when the machine and the work method were designed.

	Mechanical availability	
Feller	98%	0.98
delimber	93%	0.93
debarker	95%	0.95
bucker	98%	0.98

Total availability (efficiency or utilization) of the system 0.98 x 0.93 x 0.95 x 0.98 = 0.83% (Silversides, 1983).

Integration of partial operations in harvesting machines

Future tree harvesting operations, particularly with respect to young stands, might increasingly be integrated into entirely mechanized systems. Evolution points in that direction.

Thus, there are reasons to expect that such modes of production may influence the costs in a favourable direction. This possibility is of particular value with respect to timber from thinning. At current development of costs the limit of profitability in thinning tends to glide towards a larger diameter.

Efficient and rational mechanization could stop this trend towards a larger diameter. It would make more timber available, promoting stand treatments. The inferior trees would be removed by thinning and growth of the best trees is then accelerated. Trees remaining after the operations would be grown more rapidly into large sizes which in turn can be harvested at lower future costs.

Current development and coordination of various operations is supported by the following mechanization motto: Optimum modes of production should be attainable by means of continuous work operations carried out in motion. Integrated partial processing in the operations is coupled parallel to a simultaneous process which also allows a rate of production flow that varies with the tree size. (Staaf, 1972).

The various partial operations can obviously be combined, coordinated and integrated by the use of machines which can do the felling (harvesters) and/or other operations (processors). The extent of processing into various sizes of timber and the ensuing methods of cross-country transport thus can be put into various methods which are considered most practical and common.

Transports of timber in terrain

Transports of timber in forest operations are a dominating part of the total work input. These transports from the sites of the trees to various points of manufacture in Sweden amount in some instances to over 2000 million tonne-kilometres per year, or equivalent to transporting 5000 tonnes 10 times around the earth.

Cost of transports

It is of great importance for the viability of the forest operations that transports are carried out at the lowest cost possible. To succeed in this respect, it is necessary to know thoroughly all the factors that influence transport output, utilization of the means of transport available, and the structure of the partial transport costs.

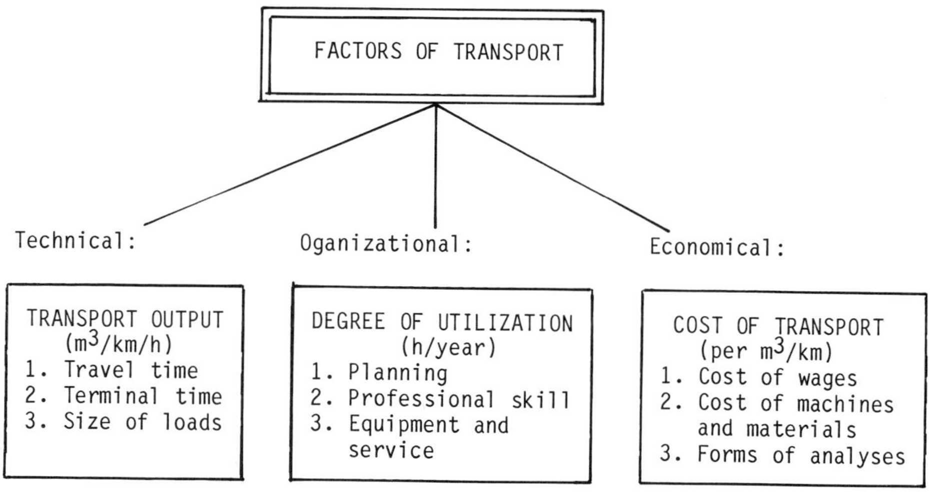

FIGURE 140. Factors influencing transports (Staaf, 1972).

The most important factors of transport are identified in Figure 141.

Most factors that influence the transports can be treated or influenced by various measures in order to minimize the cost of transports. The cost of transports is the ratio of the annual cost of all timber transports in a region and the volume of timber transported, a relationship which can be expressed by the following formula:

$$\frac{\text{Annual cost}}{\text{Annual transports}} = \text{cost per unit}$$

The formula can be converted into the following general expression by means of some of the transport factors discussed above:

$$\frac{\text{Capital cost + operational cost}}{\text{Transport output (m}^3\text{/h) x utilization (h/year)}} = \text{Cost per unit}$$

Every factor that has an influence on the cost per unit in the formula above will be analyzed and discussed in a subsequent section. The possibilities to make the values above the line as low as possible and the values below the line as high as possible will then be subject to special attention (Haarla, 1973).

Some transport concepts

Transport is usually a transfer of an object from one point to the other. This transfer is often a link in a chain of major production events. Input made into the transport is expected to be entirely counter-balanced by the output from the transfer.

Work involved in transports of timber can be defined as shipment or transfer of timber from one point to another and to services diectly or indirectly associated with the shipment or transfer.

Transport in terrain and transport on roads

A number of various transport terms are used in forest operations. The concepts of forwarding and further transport have previously been presented under the heading "Tree harvesting terminology".

However, the term transport in terrain (off-road) which has been defined as a transport of timber in terrain without roads and on very simple often temporarily built roads, is not analogous to the term forwarding. The term transport in terrain is associated with the term transport on roads which is transfer of timber on roads as well.

Short transports and long transports

With respect to transport distance it is customary to distinguish between short transports and long transports, short transports being transfers of timber from landings over short distances to points of marketing while long transports are transfers over long distance to points of marketing.

Driving and terminal work

A transport operation, e.g. a transport cycle consisting of shipment and return of equal distance, can be divided into two different forms of work, viz. driving and terminal work.

Driving is defined as work which depends on the characteristics of the transport route, distance etc. Terminal work is tranport activities which are independent of the characteristics of the transport route.

Transport of load and empty transport are two distinct phases of driving.

Loading and unloading are two different phases of the terminal work. See Figure 141 showing the four phases of the transport cycle.

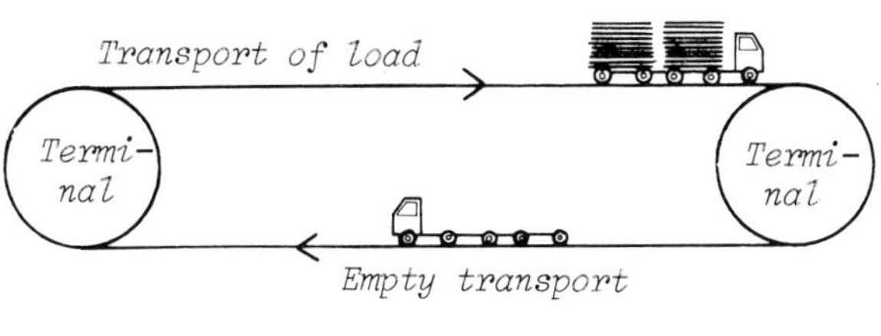

FIGURE 141. The four phases of the transport cycle: empty transport, loading at terminal, transport of load and unloading at terminal.

Forest roads and timber terminals

Most common types of forest roads

Transports in forest operations can be carried out on various transport routes: on land, water and in air. This will be discussed further in a subsequent section. Only definitions of our most common types of roads built for transports of products from the forests will be given here.

Access road is a forest road which has made an area accessible. Primarily it has an administrative function i.e. the benefits of the road are largely transports of goods and labour to and from the forest area.

Collector road, which is connected to the access road, is often the outer branch of a road system mainly with crosswise function, i.e. the benefit of the road is largely transports of personnel and goods from and to the surrounding area.

Strip road is a temporarily used collector road, usually with a dead end in terrain. Depending on the type of transport and vehicle used it can be termed skid road, forwarding road, horse strip road or tractor strip road.

Terminal locations

As in the case with driving, which is carried out on various types of roads or transport routes, terminal work is done in various locations.

Timber terminal is a major storage place where timber is stored temporarily for processing and further transport.

Storage place is an area set aside in the forest for concentration and storage of timber, other forest products or road maintenance equipment.

Landing or place of piling is an area where timber is concentrated for further transport. This is usually also a place for change of transport mode. Landing may also be a place on a river bank above the highest water mark.

Ice landing is a landing on floating ice;

Forwarding

As mentioned previously transports in forestry operations are an essential part of the total harvesting system. It has also been shown how some phases of transport are included in operations not normally considered to be parts of transport.

Felling, processing and bunching of timber interface with transport. For instance, directed felling is of great importance for the continued handling as is bunching of timber along strip roads.

This first collection of timber from individual trees and trunks over a large area of felling has a low output and high costs of transport.

Actually, transport of timber starts in swaths or routes specially prepared for forwarding and by means of vehicles designed for this purpose. The first phase of forwarding is predominantly a transport in terrain.

Choice of transport method in terrain

The choice of transport method in terrain depends on conditions such as type of timber, terrain and means of transport available. Of special interest is the hauling power of the equipment.

The horse

Animals, primarily the horses, have long been used for hauling of timber in forest operations. Oxen have also been used. For a full utilization of the rather limited hauling power of the horse, between 1000 N and 3000 N (100 kp - 300 kp), it was necessary to set high standards on equipment (harness and horse shoes), the design of the carrier, and the road surface.

It is important for personnel responsible for the hauling of timber by means of limited resources to facilitate the operation by arranging for efficient traction power and to avoid all losses of efficiency caused by erratic planning, excessive distances, adverse slopes, unfeasible equipment design etc.

Measures that can be taken in order to achieve improved traction when horses are used include the development of horse shoes with spikes, cleats or hooks that correspond to the pattern of tractor tires, anti-slip devices, chains or tracks mounted on pulling wheels.

Friction at horse transports has been reduced by the use of sleds developed for roads on snow or ice and wheel carriers for bare ground conditions.

Expanding truck road systems in the forests

After the second World War the development of better crawler tractors promoted mechanized road building techniques in the forests and brought about a rapidly increasing rate of road construction. The expanding truck road systems reduced the transport distances in terrain. The average distance from the stump to the truck road was reduced to 1 km in northern regions and to about 0.2-0.4 km in southern regions. The average transport distance in terrain is now approximately 0.5-0.7 km in many countries.

From the beginning of the 1950's the shorter transport distances in terrain enhanced the opportunities for bare ground transports. Due to denser road systems, the previous difficulties in terrain such as sites with low carrying capacity, bogs and swamps, became less important obstacles for bare ground transports by horses or tractors. In addition, e.g. reduced availability of horses and labour in the forested regions made it opportune to introduce year-round harvest operations.

Tractors for tree harvesting

In the middle of the 1950's small (crawler or semi-crawler types) farm tractors were used for transports of timber in terrain.

In the 1960's tree harvesting tractors equipped with hydraulic grapple loaders which facilitated loading and unloading operations were introduced. This development reduced further the number of horses.

Current transport infrastructure

Various objects of transport, transport routes with terminals and means of transport will now be discussed in order to elucidate the current transport situation.

Objects of transport

Volume, weight and shape of timber

Timber as an object of transport is discussed in the following section with respect to quantities and sizes of the individual objects of transport, characteristics of piling and other features i.e. volume, weight and shape of timber.

Quantity of transports

As an example it may be mentioned that the total annual quantity of timber felled and transported in a northern country may currently be approximately 70 million m^3. This volume is largely collected from approximately 50 percent of the total land area for transport to a limited number of wood processing industries.

Volume of timber

About 70 percent of the total annual quantity, or approximately 50 million m^3 is obtained from final harvest areas and 30 percent or 20 million m^3 from thinned stands.

If an average rotation period of 100 years is assumed, the area clearcut each year would be 1/100th of 23 million hectares, or 230 000 hectares. This area yields approximately 50 million m^3 with an average of (50 000 000 : 230 000) = 217 m^3 per hectare.

The amount of timber from thinned stands depends on i.a. the number of thinning operations carried out during the life of the stands. If it is assumed that the stands are thinned once, the average amount of timber removed per hectare would be (20 000 000 : 230 000) or 85 m^3.

If two thinning operations are carried out during the life of the stands, approximately 2 percent of the forest area would be thinned each year, yielding 20 million m^3 from 460 000 hectares, which corresponds to an average of 43 m^3 per hectare.

The average quantity of timber per hectare in Sweden may amount to:

- In final harvest stands 200 - 250 m^3
- In thinned stands 40 - 90 m^3

The amount of timber per hectare and the area harvested give the total quantity of transport from the operation. This quantity influences the cost of tree harvesting to the effect that the costs decline at increasing volume of timber within a given area - to a certain limit. This applies in particular to mechanized harvesting systems.

Costs of tree harvesting

At clearcutting, the costs of tree harvesting remain constant for areas larger than 6 - 7 hectares.

At thinning a much larger area must be treated in order to obtain a total quantity of, say, 1000 - 1500 m^3. When an area of 20 hectares and larger is thinned the cost of harvesting seems to stabilize. This also applies to the cost of establishing a stand (Staaf, 1953).

Since the area of a stand is usually constant during the various stages of regeneration, thinning and final harvest, it is important in the long term planning to consider the total area dependent costs of the various measures.

Size of the clearcut areas

The average size of clearcut areas is less than 10 hectares. The clearcut areas are larger in northern regions than in southern regions and larger in industrial forest operations than in farm forests.

Dimensions of the transport objects

The dimensions of the average tree as well as the volume of timber per hectare influence the output and the costs of transports, the latter declining at increasing average tree diameter at breast height.

The average tree diameter varies with species, age of the trees and site quality and it is influenced by various forms of stand treatments. Thus, thinning from below (low thinning) yields small timber while thinning from above (high thinning) yields big timber.

Various objectives in the forest operations may also produce differences in the dimensions of timber. Compare industrial forest operations with specialized forest operations in production of e.g. sawtimber, pilings and poles etc.

Various dimensions of trees produce different units of handling such as chips, bunches or piles of standard length or random length, tree length trunks, pieces of trunks, tree bolts or sawlogs. This has been discussed in a previous section.

Sorting of timber influences the cost of transport in terrain. The more assortments, the longer time is required to obtain a full load.

Piling of the transport objects

Piling, or popularly expressed serving of timber for transport, is an important step taken in order to facilitate a high transport output.

Bunching and piling of timber should take into consideration the direction of transport, the accessibility of timber and the loading equipment. This applies in stands, at strip roads and landings, thus, at the various terminals.

The methods of piling or the forms of bunching vary between the systems of harvesting, which has been discussed in a previous section.

Weight of timber

Other characteristics of the timber influencing the transport work are e.g. various degrees of debarking and dryness, and the absence of slash, debris, soil, snow, ice etc. that contribute to the weight of timber.

Timber, debarked and dry, is 20 - 25 percent lower in weight than recently felled timber. If the timber is also free of ice, snow, slash etc. transport work can be done more efficiently and at lower costs than if the timber is recently felled, soiled and poorly piled.

However, a reduction in the cost of a transport operation may be achieved at the expense of a subsequent handling and processing operation during the movement of timber from stump to the mill.

Routes of transport

The following chapter will deal with various forms of forwarding. Transports in terrain will be discussed with particular attention to roads i.e. patterns, distances, maintenance and terminals.

Various routes of forwarding

Collector roads and strip roads are normal routes of forwarding prior to further transports on truck roads. A functional difference between collector roads and strip roads is that the collector roads are meant to be used only for transports while strip roads will also be used for loading.

Transport of timber along skid trails by means of cables and cranes is another form of forwarding. The transport function of the cable (crane) is primarily comparable with the function of a collector road while the lifting cables and the skid trails are comparable with the strip roads (Samset, 1981).

Floating of timber in creeks (small rivers) was previously a form of forwarding to the main water courses which were used for further transport.

Patterns of strip road systems and road net density

Routes (swaths) for transport of timber in terrain can be laid out in various patterns, primarily depending on the amount of timber, means of transport and terrain conditions.

Various patterns of road systems

When large amounts of timber are harvested per hectare as is the case in final harvest operations, bunching is efficient, expressed e.g. in m^3 per 100 m of strip road. Simultaneously, distances between the strip roads are relatively short (8 - 12 m).

When the amount of timber per hectare is small as is the case in thinning operations, distance between the strip roads must be longer, e.g. 16 - 24 m, in order to provide for a satisfactory efficiency of bunching. Since bunching over long distances is expensive, it must be weighted against the gain in having a large amount of timber at the strip roads. The gain from extensive bunching to the strip roads consists in quicker loading work.

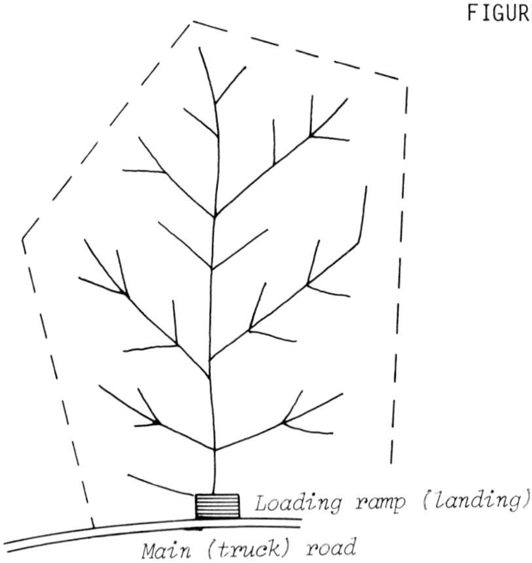

FIGURE 142. Arbor like pattern of strip road system.

At transports in terrain by horse, so-called skidding, an arbor like pattern of road system was used (Figure 142). This pattern was laid out with special attention to the limited hauling power of the horses in slopes. Loading started far out in the road system and continued en route to the loading ramp or a terminal for unloading and further transport.

At transports in terrain by means of a tractor designed for harvest operations, a different and greater hauling capacity is put into action. Transports of large loads and reduced sensitivity to adverse slopes in particular have lead to a pattern of parallel strip roads (See Figure 143).

If the load capacity and the travel speed of the modern forest tractors are utilized fully, long and straight strip roads have produced the best transport output. Side slopes, to which forest tractors are sensitive, are avoided carefully, particularly when the tractors are loaded and have a high centre of gravity.

The strip roads should be connected by means of cross-roads for so-called loop driving in order to avoid time consuming turn-arounds.

In steep slopes the road system may be laid out in a zig-zag pattern (Figure 144).

A spiral shaped road system might be feasible in a more or less conical terrain sector.

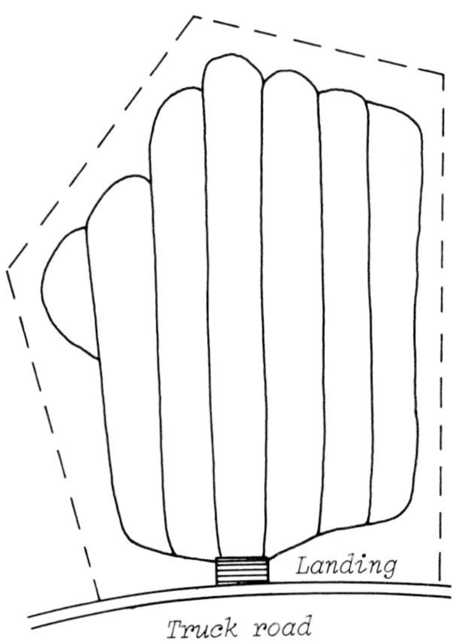

FIGURE 143. A pattern of parallel strip roads.

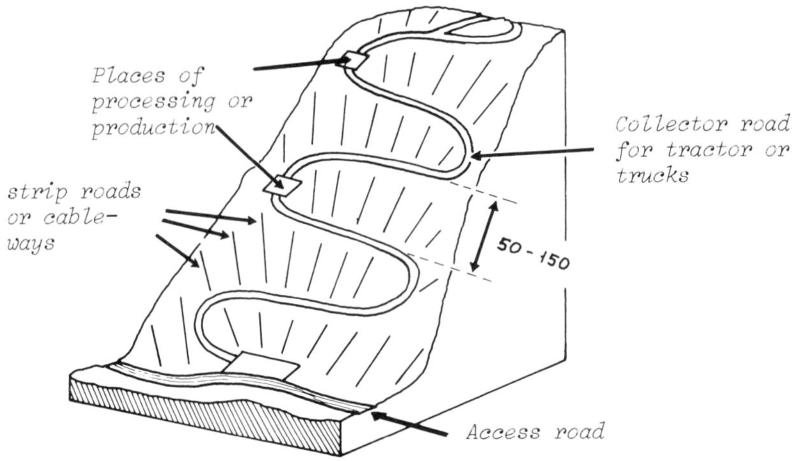

FIGURE 144. Road system in steep slopes.

Optimum density of the road system

The lay-out of a simple road system in terrain can be done by marking strip roads and collector roads with tapes of different colours. Optimum density of the total road system should then have been calculated.

Optimum density of the road system depends i.a. on the amount of timber to be harvested per hectare, size of timber, costs of road construction and maintenance, means of transport and method of harvesting. Of these factors it is primarily the amount of timber and the cost of road construction that determine the optimum density of the road system.

To calculate the optimum density of a road system for harvesting, it is often necessary to prepare alternative transport analyses. These analyses may be rather difficult to carry out because of deficient data on time and costs, primarily concerning work on the system of strip roads. Information may be scanty on various terrain conditions, carrying capacity, frequency of boulders, undergrowth, snow depth and methods of transport (Putkisto, 1958).

Length of strip roads

The rapid expansion of road systems by means of efficient road construction machines has reduced the length of strip roads needed for transports in terrain. The extension of truck roads has been based on calculations showing that the incremental costs of transport on the new roads are balanced by the cost reduction on shortened transports in terrain. Thus, the analyses have given the alternative which has the lowest total cost of transport.

A reduction of transport in terrain or on strip roads shortens the travel time for this kind of transport. The travel time is obtained from the length of the strip road and the speed of driving as follows:

$$\text{Travel time} = \frac{\text{Length of strip road}}{\text{Speed of driving}}$$

Reducing the length of strip roads by efficient road planning, the forester can increase speed of driving and influence the travel time, which is particularly important when expensive transport equipment is used. Length of strip roads can also be shortened further if the roads are built as straight as possible.

The straight line distance between places of loading and unloading is often 15-25 percent shorter than the real road distance in terrain. The discrepancy is called allowance for winding, which consequently is the difference between the actual distance of skidding and the straight line distance.

Since transports in terrain are more expensive per tonne-km than transports on roads, it is important to establish the shortest possible road distance in terrain to the truck road. The routes of transport in terrain should be laid out perpendicularly to the truck transport routes.

While the techniques used in road construction are made more efficient by development and transport of timber in terrain becomes more expensive, a gradual expansion of the forest road system should be implemented. The optimum meeting point between trucks and tree harvesting machines or forwarders should be calculated continuously. Since the modern trucks have developed into transport means of great capacity and efficiency, they should also be utilized to the greatest extent possible. In some cases the felling machines may operate as an excavator by a placement of the felling head with a suitably designed shovel for trenching and grading of the extension of a truck road. This has been practiced with very good results technically and economically (Staaf, 1983).

FIGURE 145. Three types of base machines for tree harvesting and timber transport in terrain.

Relation ship between strip roads and the truck road

The basic problem is how close to the stump should the truck road be extended and how dense should be the system of strip roads. It is assumed that the terminal costs and the general transport costs (management, supervision, buildings etc.) are not influenced but constant for various densities of road systems and that the amount of timber is evenly distributed within the area influenced by the roads. The following concise objective can then be established: "definition of the meeting point between the two phases of transport where the sum of the direct and the indirect travel costs of the phases of transport of timber from the point of origin to the final destination is at a minimum".

The lowest total cost (K) of direct (D) and indirect (I) travel costs (U) for both transport in terrain (K_1) and transport of truck roads (K_2) is then calculated according to the following formula (Sundberg, 1952, 1953):

$$K = DUK_1 + IUK_1 + DUK_2 + IUK_2$$

The calculations should indicate that the density of the truck road system and, hence, the length of strip roads depends on the cost of the truck road construction, the specific travel cost of transport in terrain, and the quantity of timber per area unit to the effect that high costs of truck road construction lead to increased distance between the strip roads and that high travel costs in terrain and/or large amount of timber will reduce the optimum distance between the strip roads.

Terminals

A timber terminal is a storage place or a landing where timber is collected for processing and/or reloading for further transport (See Table 7).

Work at the terminals with loading and unloading and work with the direct transport between the terminals is the total work content of a transport cycle, popularly called a turn or round-trip.

Lay-out and denotation of the various terminal places depends on the function and position of the terminals in the chain of transports.

Further down the chain of transports the terminals usually increase in size since volume of timber increases.

Various types of terminals

A terminal often serves as a place of delivery. Thus, timber can be delivered at truck roads, water cources, railways or at industrial sites.

Efficient reloading and processing operations at the terminals require careful planning of the work procedures which must be well thought through, perhaps after analyses of time and methods.

Table 7. Various terminal places

Transport terminals		
Land terminals	Forest terminals	Buffer terminals
		Storage terminals
	Shore terminals	High piles
		Low piles
Lake terminals	Terminals on water	
	Terminals on ice	Terminals snow cleared
		Terminals with prepared ice
Industrial terminals		
Terminals on land	Buffer terminals	
Terminals on water	Storage terminals	

Terminals on ice

This type of terminal may be of special interest in northern regions since it can be arranged artificially by preparation of the ice surface. Terminals with prepared ice may have the following advantages:

- the place may be spacious
- extended winter operation

- improved unloading and loading
- improved concentration of timber
- improved storage of timber and accelerated drying
- more rational scaling
- more efficient processing by means of machines
- improved safety

Preparation of ice

Preparations are done to improve the formation of ice in order to obtain strong ice of a certain thickness as soon as possible (Ager, 1963).

The preparation of ice is usually done by snow compaction and watering. Preparation is initiated early in the winter season in order to obtain an ice thickness of at least 75 cm necessary for truck transport.

Various methods of ice preparation

Ice with snow cover can be treated in various ways e.g. watering only, snow compaction and watering in combination, snow compaction only, or snow

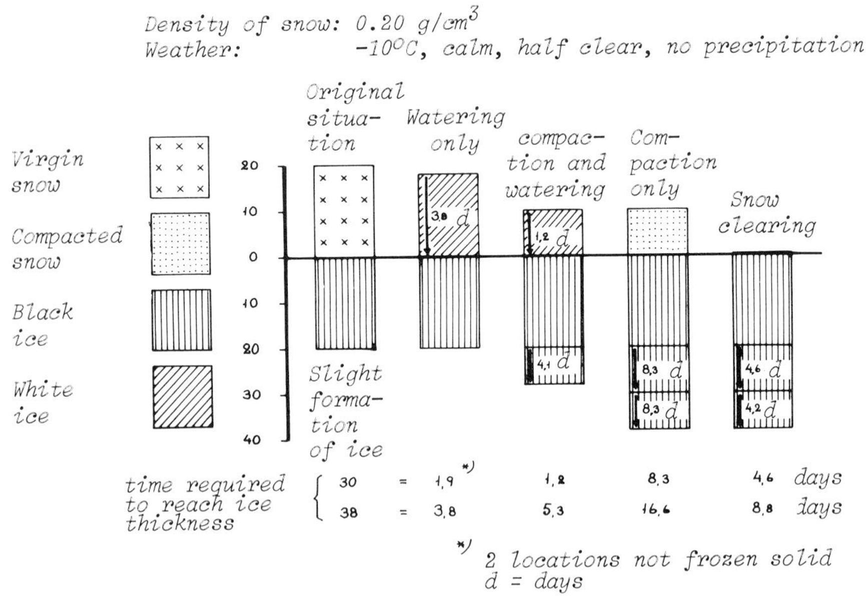

FIGURE 146. Rates of ice formation for various methods of preparation (Ager, 1963).

removal. The rates of ice formation at a temperature of -10°C and for various methods of preparation are shown in Figure 146.

Quality of ice

Whether truck traffic on ice is possible depends not only on the thickness of ice but also on quality and other characteristics of the ice. Resistance of the ice to breakage is of great importance. Black ice has greater resistance than white ice.

Repeated watering creates a layering of the ice that prevents deep penetration of cracks. This measure reduces the risk of loads sinking through the ice which may happen under certain circumstances even on 1-m thick ice with a special pattern of crack formation, e.g. restricted triangular pattern. Timber transports on ice are always risky.

The carrying capacity of ice is relatively low. It depends on the density of ice which is approximately 0.90-0.92 g per cm^3 for black ice and 0.88-0.91 g per cm^3 for white ice. The carrying capacity is obviously slightly higher for white ice.

How is watering done?

Preparation of ice is normally done by means of motor pumps when the thickness of ice indicates a certain carrying capacity and when snow cover is sufficient. Pump units on the market have capacities varying between 3000 litres and over 10 000 litres per minute.

Means of transport

Means of transport are primarily vehicles normally consisting of a power unit and equipment to carry the load. There are two types of power to consider in this context, viz.

- muscular power
- mechanical power

Means of transport on land, water and in air

Means of transport are designed for operations on various routes and under different environmental conditions. A distinction can be made between means of transport for operations on land, water or in air.

Transports on land

Transport of timber on land is almost always done by trucks (lorries), tractors etc. on the ground surface. Transports on land can also be done by means of track carriers. However, timber transports by rail are not common.

Transports on water

Transports of timber on water still occur on some rivers and at sea rafting. Tug boats of various sizes are then used for hauling of timber, loose or in bunches, over stretches where the currents are not strong enough to keep the timber moving at sufficient speed.

In rivers with strong currents, water serves as a means of transport due to the gravitational pull as a source of power and the water is a load carrying medium.

Transports below the water surface

Transports of timber below the water surface can be done in systems of pipes with a large diameter where chips can be transported in a slurry consisting of approximately 50 percent water and 50 percent chips. The mixture of chips and water is moved through the pipe system using pump stations as the source of power.

Transport by aircraft

Means of transport in air are increasingly being used. Small aircraft and helicopters are used in forest operations for miscellaneous transports and other activities such as fire control, fertilization, air photography, reconnaissance, moose census etc (Samset, 1972).

Attempts have also been made at developing transport techniques for timber by means of large helicopters and balloons in difficult and remote terrain.

Live means of transport

Of muscular power sources for transports may be mentioned, in addition to Man, animals such as horses, oxen, mules, reindeer, camels and elefants. An estimated 450 million animals of various species are occupied in transports of products throughout the world. Horses have been the most important animals used for transport of timber in the Nordic countries.

While horses have become almost extinct in large scale tree harvesting operations in recent years because of the rapid progress of mechanization, the use of horses in small forests (woodlots) has gained recognition (Hedman, 1983).

A suitable work horse may not always be a cheap investment but it will require no expensive oil or gasoline for operation and the appliances needed may be relatively simple and cheap.

There are also a number of harvesting situations in which the use of a horse (animal) is advantageous e.g.

- in small areas of final harvest (clearcutting or selective felling)
- in thinning operations where the net of strip roads is very open
- for harvest of seed trees and windthrown timber

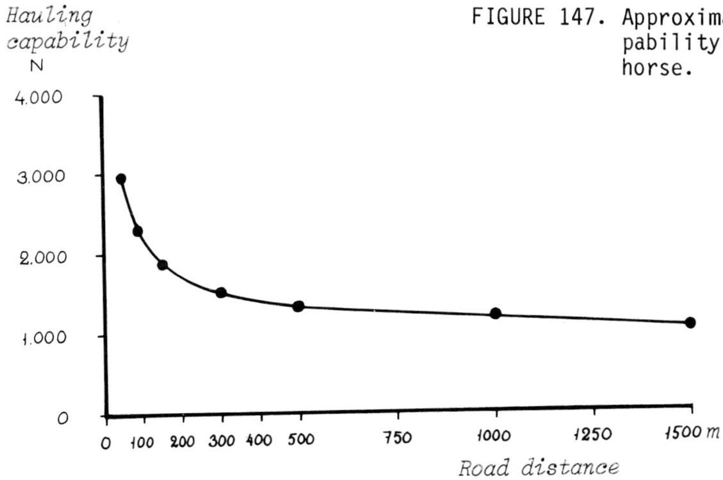

FIGURE 147. Approximate hauling capability of a normal horse.

- in difficult terrain
- in cleaning and clean-up operations for removal of fuelwood

Conditions for transport by horse in general

At transport there are primarily two factors that should be considered closely with respect to the utilization of a given source of live power. To derive a maximum possible result of practical haulage, every measure should be taken in order to get efficient traction and to ensure that the transport will be carried out with a minimum possible loss of power, e.g. by using well designed load carrying equipment with low friction.

The performance of a horse in haulage

A horse can sustain a hauling force of approximately 1000 N (Figure 147). Tests can give information on the hauling capabilities of individual horses with respect to their power, energy and techniques.

Traction

To achieve efficient traction the horse is equipped with special shoes having various sharp devices or cleats.

Minimum possible loss of power

For a minimum loss of power at transport by horses, it is important i.e. that the harness and the pulling arrangements (shafts) are well adjusted (Kubiak, 1976). It is important also that the vehicle has a minimum of

FIGURE 148. A skidding cart with wheel base.

friction e.g. resistance to gliding of runners and to rolling of wheels, and that suitable down-hill slopes are utilized as much as possible.

At transports on bare ground, a large proportion of the hauling capability of the horse must be utilized in order to compensate for friction. Wheels, therefore, were mounted on equipment used for transport by horse already hundreds of years ago (Staaf, 1963) (Figure 148).

Table 8. Resistance to rolling on bare ground, vehicle with 4 rubber tire wheels (VSA).

Type of friction on bare ground	Equipment	Weight kg	Resistance to rolling in percent of vehicle weight					
			Forest land		Other land			
			Pine heath sand	Till w/o. boulders	Till with some boulders	Till with many boulders	Wet meadow	Gravel road
Resistance to rolling	Skidding cart w. 4 rubber wheels, convex surface	200	19-21	19-22	29-30	40	27-28	13
Resistance to gliding	Skidding sleigh with runners 90 mm wide	80	28	32	38	42	33	47

Table 9. Resistance to gliding at transport on snow. Resistance to friction in percent of the total weight of vehicle. Conditions: Horizontal surface and snow conditions typical of northern regions. (SDA).

Equipment and load	Resistance to friction in percent of total weight				Temperature
	Loose snow	Compact snow track	Winter road prepared	Winter road with ice tracks	
Sleigh with load lifted from ground	5-15	2-10	3-5	0.3 - 2.5	-8°C
Sleigh with rough timber skidding	25-35	20-24			-20-22°C

Tables 8 and 9 show the high resistance to gliding of equipment with runners on bare ground compared with that in show and the considerably lower resistance to rolling of wheeled vehicles on normal bare ground compared with the resistance to gliding of vehicle with runners on bare ground.

FIGURE 149. Forwarding birch pulpwood by horse.

The following harvesting situations may be described for transport of timber by horse.

Final harvest operations

It may be advantageous, particularly for self-employed forest owners, to use a horse in the harvest of mature trees from small areas or from projects of short duration where a heavy investment in machine equipment is unfeasible or prohibitive. The use of horses may also be the only alternative for usually short transport of timber in difficult terrain or on sites that are soft or covered with valuable seedling stands.

Equipment used for transport by horses in final harvest operations should have a capacity of 2 m^3 - 5 m^3. Because of the limited hauling capability of horses relative to that of tractors, it is imperative that resistance to haulage (friction) is reduced. While sleds can be used in cold seasons with frost, ice or snow covered ground, carts or wagons with wheels is the proper equipment on bare ground. It is also important that equipment for transport by horses is designed to facilitate loading and unloading.

Thinning operations

The net of wide strip roads can be reduced considerably when horses are used in thinning operations. Damage to roots and trunks of remaining trees will also be less. The following examples show how a horse may be used in two different thinning operations.

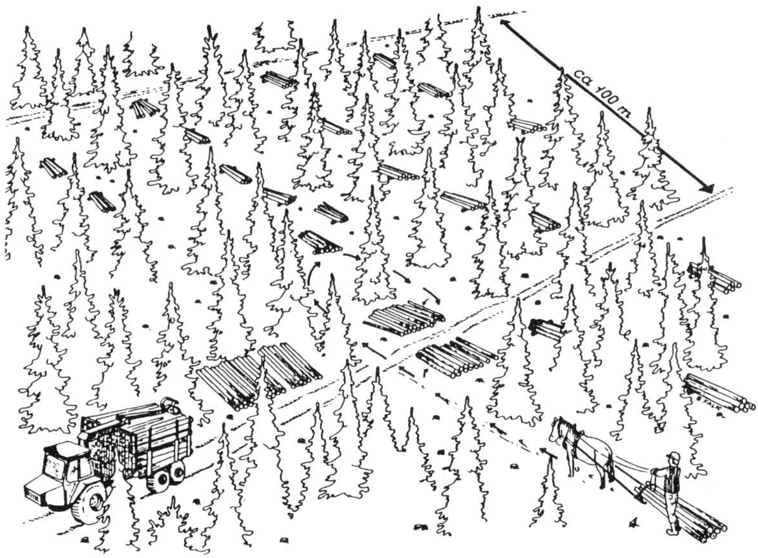

FIGURE 150. Horse skidding to strip roads. The timber is then reloaded to a forwarder for transport to landing.

The transport output achieved at horse skidding in thinning operations (Figure 150) may amount to about 7 m^3 per hour when the average distance between the strip roads is 100 m (News for Small Scale Forestry, 1982). Transport of the logs directly from the stumps to the truck road requires great load capacity and low friction.

Loading equipment should be efficient, making work at landings less strenuous. Output at horse skidding in thinning operations is also influenced by the distance of transport (Figure 152).

FIGURE 151. Transport by horse from the stumps to the truck road.

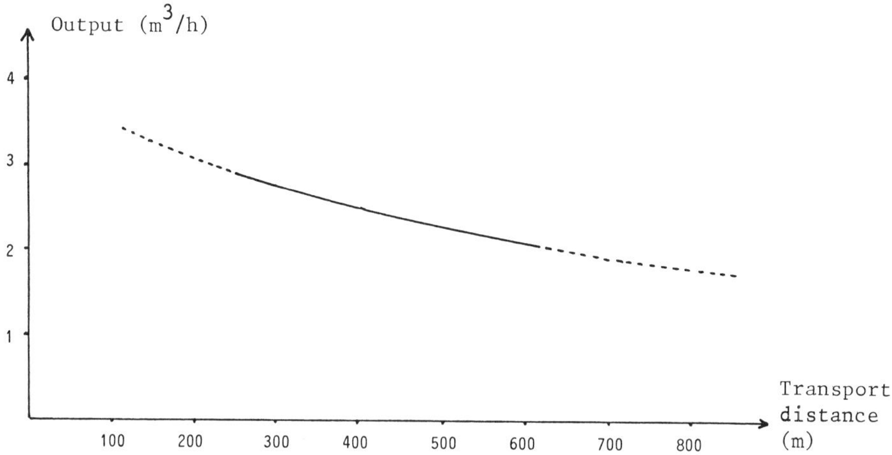

FIGURE 152. Output capacity at transport by horse (double sled) to landing (Hedman, 1983).

Other tree harvesting operations

The horse can be used advantageously when harvesting seed trees and scattered windthrown trees. Removal of a small volume of timber by means of high performance, expensive machine equipment is often more costly than removal by means of slower and relatively cheap equipment.

Harvest of fuelwood from cleaning operations

The horse can also be used for harvest of fuelwood from cleaning or clean-up operations.

FIGURE 153. Skidder equipment for horse.

- Width: 100 cm
- Bunk clearance: 22 cm
- Weight: 42 kg
- Load capacity, approx.: 0.5 m^3 (solid wood).

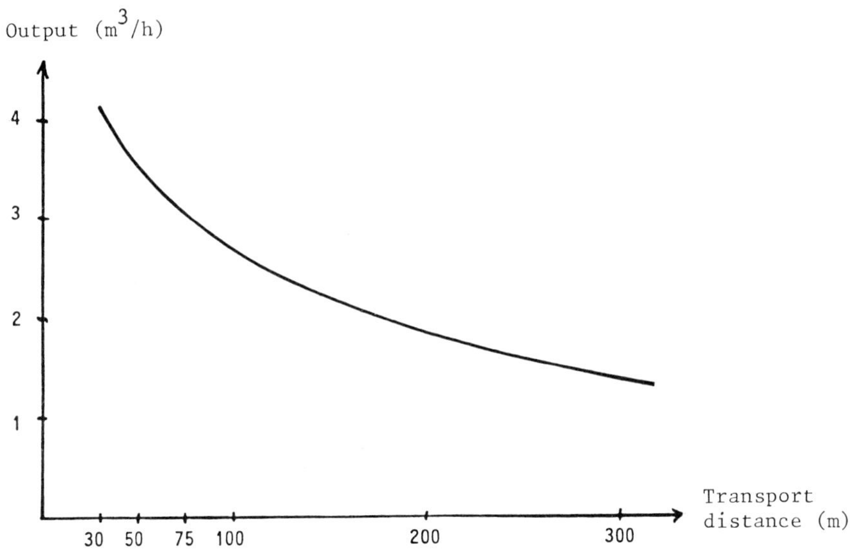

FIGURE 154. Output capacity at skidding of whole trees from a cleaning operation.

FIGURE 155. Bogie trailer with hydraulic equipment for loading.

Some types of equipment used at transport of timber by horse (Hedman, 1983):

FIGURE 156. a) Skidding pan of glass fiber. b) Skidding pan of light metal

Length:	90 cm - 120 cm
Width:	100 cm - 105 cm
Height:	40 cm
Weight:	18 kg - 24 kg

Carts (wagons) for transport by horse on bare ground (single axle or bogie)

FIGURE 157. Wagon for transport of timber by horse on bare ground.

Weight:	200 kg - 300 kg
Wheels:	7.70 - 10
Load height:	58 cm - 68 cm
Load capacity:	3 tonnes

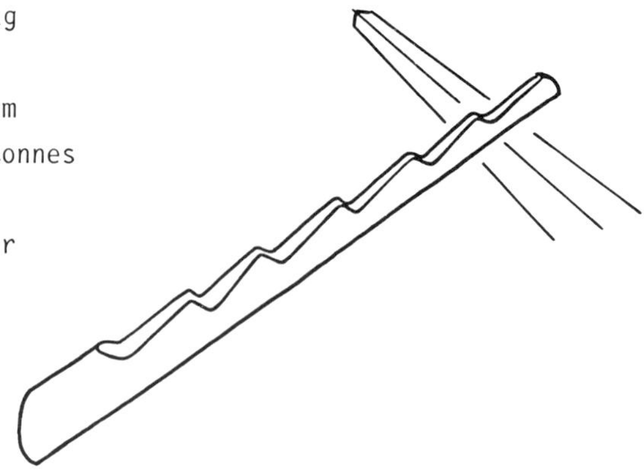

FIGURE 158. Simple log jack for manual loading of heavy timber.

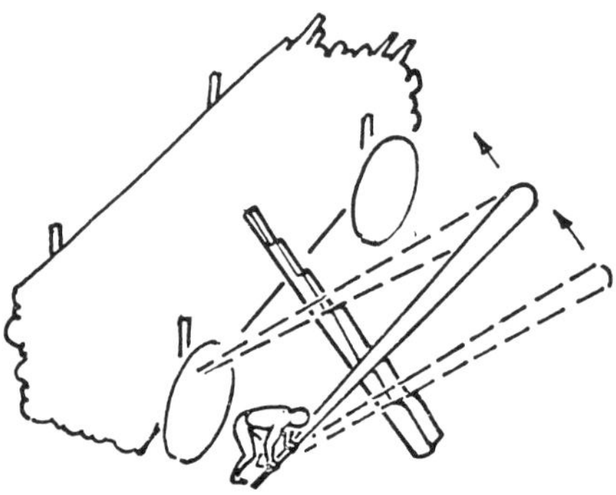

FIGURE 159. Manual loading of timber by means of one log jack.

FIGURE 160. Manual loading of timber by means of two log jacks.

Manual winches

Several types of manual winches range in hauling capability between 200 kg and 1200 kg for 1 - 2 gears.

FIGURE 161. Manual winch.

FIGURE 162. Motor powered winch.

Motor powered winches

Hauling capability:	400 kp
Weight:	16 kg
Cable speed:	0.3 m/s
Cable gauge:	4 mm
Cable length:	30 m

Tractor as a means of transport in harvesting operations

Of the machines and mechanical means of transport available for forwarding today primarily the tractor with loading equipment and some winch arrangements will be presented in this and the following chapter.

Development of the tractor

In Sweden tractors were introduced into forest operations as a means of transport by farmers in the beginning of the 1950's.

Tractors have also been used in the forests for purposes other than transports i.a. for work on construction of roads and landings, snow removal, piling of timber, timber handling at terminal places and in forest improvement (Figure 164).

The great automobile producer Henry Ford, who manufactured no less than 15 million T-Fords in the beginning of the 1900's also contributed strongly to the development of the farm tractors.

The forest tractor

As the use of horses declined, the lost hauling power was replaced in forest operations with tractors initially equipped with i.a. semi-tracks over the rear wheels and two supporting wheels underneath the middle of the tractor (Figure 163).

FIGURE 163. Farm tractor equipped with semi-tracks for forest operations - in this case a forwarder transporting the timber entirely above the ground. It is a tractor hauling a trailer as carrier. (Lundaahl, 1961).

Initially the tractor was not so efficient and economic as the horse. However, the technical development and special requests for a tractor that could be used as a means of transport combined to produce eventually a true forest tractor. Several types were equipped with big wheels, four-wheel drive and new steering mechanisms e.g. frame steering.

Requirements of the tractor

In view of what is required of a forest tractor, the objective of development has been to achieve the highest possible transport output (m^3/h) and a high degree of utilization (h/year) at the lowest possible cost. For the achievement of a high transport output is required a machine with great ability to travel in terrain, a high hauling capacity and efficient equipment for loading and unloading.

FIGURE 164. Four-wheel drive forest tractor equipped with a winch for skidding of timber. It is here used for bunching of tree length trunks at landings (Staaf, 1964).

Ability to travel in terrain

The requirements concerning ability of the forest tractor to travel in terrain applies in particular to transports with full loads in terrain and on roads.

The tractor wheel

The transition from horses and vehicles with runners, primarily for transports on snow, to tractors has made the wheels of the tractor and its load carrying vehicle very important machine components worth serious attention.

Improved knowledge of wheels and wheel combinations is needed

Improved knowledge of the optimum design and sizes of wheels is essential for the development of mobile machines for processing, handling and load carrying equipment in forest operations. This applies to various combinations of wheels as well. The rapid change towards larger sizes of wheels in the 1960's, a development which was largely intuitive, demonstrated the need for improved knowledge in this important sector of transport technology. Some of these matters have been subject to research by means of laboratory experiments and full scale field testing. Thus, the behaviour of wheels and wheel carriers when passing over individual solid obstacles such as rocks, stumps etc, has been studied.

A large number of data have been obtained for the purpose of comparison when wheels of various sizes, designs and tire pressure have been tested. The behaviour of the pulled wheel depends not only on the individual characteristics of the wheel but also on its suspension.

Difference in resistance to rolling between twin wheels and single wheels

Comparisons concerning traction of various tire designs of unloaded wheels show e.g. that twin mounted truck tires require approximately 13-18 percent more traction power at a tire pressure of 35 N per cm^2 than that required for single mounted tires of (LP) low profile type at a tire pressure of 40 N per cm^2 (Figure 165).

FIGURE 165. Low profile tire (A) and twin mounted truck tires (B).

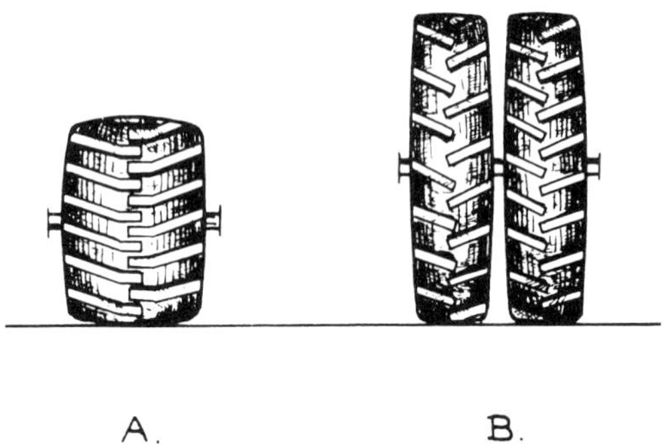

Research is being carried out concerning the proper size and design of the individually pulled wheel on forest land with variable "elastic" characteristics.

There are today rather well designed wheels and tracks that facilitate driving in terrain. Situation is not so good with respect to wheels for travelling on rocky, swampy or soft ground where it is necessary to have wheels that float high on the surface and can "absorb" the ground obstacles.

Standardization of tractor wheels

At present there are on the market a large number of wheel tires of various designs and sizes. The wide range of tires means higher costs because of short series and low utilization of the tires. Time appears to be ripe for a standardization of the tractor wheels.

Many types of wheels used in forest operations are more or less modified wheels from farm equipment. However, wheels used for ploughing and straight-driving in even farm fields are not very suitable on winding strip roads or in a field of stumps.

Importance of proper design and size of the wheels used in forest operations and the need for proper wheel combinations with anti-slip devices or tracks grows with increasing degree of mechanization and with the use of heavy machines. The requirements are many, causing a combination of complex problems of biological and technical nature (Staaf, 1962a).

The wheel is the cause of biological concern

Biological problems are created primarily because of damages to the roots, trunks and ground with ensuing losses of timber and yield . Technical matters concern the ability of the machines to travel in various types of terrain with special problems concerning carrying capacity and wear.

The extent of damages that occur when heavy machines are used in thinning operatons have been investigated.

If the operation of machines is kept approximately 0.7 - 1.0 m from the trunk only occasional fungal damage of economic importance may occur. In practice this means that the strip roads should be at least 4.0 m wide if width of the tractor is 2.5 m. A wider strip road may be required in difficult terrain and for winding roads (Nilsson, Hyppel, 1968).

Damages to the ground

Results of investigations of damages to the ground are very important. The results show that we may drive a tractor in the forest if width of the strip road is adapted. However, very little is known of the differences in the nature of damages caused by the operations. The extent of damages to the ground differs between various types of vehicles and modes of transport, carried and lifted loads. Thus, the distribution of weight is entire ly different when e.g. twin-coupled tires and bogie mounted tires are being used. Distribution of pressure underneath tracks is different than that under wheels (Staaf, 1969).

Damages to the ground with ensuing reduction in growth of the remaining trees adjacent to the strip roads are caused by heavy rutting. Investigations have shown how rutting occurs when vehicles or machines are operated on soft ground of sandy, loamy-sandy and loamy till. The ruts are deepened for every trip or turn until a critical depth of approximately 10 cm is reached.

At that level the wheels get in contact with the root systems of the adjacent remaining trees.

How can rutting be counteracted?

Rutting can be prevented for the first turns if a layer of limbs is left on the strip road. However, as soon as the limbs are broken and the fine branches are worn, the tracks or the wheels sink to a depth which may be 30 cm or more after the seventh turn. As a rule, drier and more compact soil is then reached, but many roots have already been cut off or damaged.

In this situation it will be necessary to consider several factors such as the relationships between the amount of timber and the density of the strip road system on one side, and the size of loads and numbers of turns on the other side. The problem consists in optimizing the combination of these factors in order to minimize rutting caused by vehicles with various sizes and designs of wheels or tracks (Figure 166). Proper planning - no rutting (Eriksson, 1981).

FIGURE 166. Rutting occurs after repeated driving of a tractor on soft ground. It is important to plan transports in terrain so that rutting is reduced to a minimum.

Improvement of traction

Efficient traction of the pulling wheels is required for trouble free operation of the vehicle in terrain. If ground is firm, there are usually no problems with traction, the pulling power required being easily achieved due to the load capacity available at transports of timber. Thére are good possibilities to improve traction in wet places by means of anti-slip devices and by putting a layer of limbs and branches on the ground.

Minimizing losses of motor power

How to minimize the losses of motor power at transports of timber is a matter of some concern. The losses may be very great in operations with vehicles which are not designed for winding roads in terrain, varying carrying capacity of the ground and rough surfaces.

In an elastic wheel tire a large part of energy absorbed at compression is recovered when the tire expands backwards. In tracks energy is lost by friction at the expense of the pulling power available. The goal should be to achieve the least possible resistance to rolling.

On strip roads or in stands where the tractor must by-pass trees, boulders and other obstacles, the losses of power vary for different types of steering mechanism. A large part of the pulling power is used for steering an old crawler tractor by means of a so-called non-regenerative steering system such as brake steering.

A re-generative steering system is more efficient and power saving but it is expensive and complicated.

The articulated frame steering principle applied in recent years has meant a considerably improved saving of power and travel capability in terrain. By means of this desgin the machine is able to move like a snake. Additionally improved ability to travel in terrain is expected when hydrostatic operation and steering has been developed.

Forces acting around a wheel

Another technical matter of interest is the magnitude and the effects of the radial and tangential forces acting around a rolling wheel. There is also the axial interaction of forces in, underneath and on the sides of the tires when a vehicle is operated on forest land.

What is to be gained by larger wheel diameter and wider tires?

A larger wheel diameter and wider tires produce a lower specific ground pressure and correspondingly lower tire pressure. A more even distribution of pressure, less risk of punctures, improved absorption of ground obstacles, a reduced resistance to rolling with less tire wear and lower costs are also experienced (Figure 167).

Research and development is expected to produce suitably sized, flexible and puncture proof tires for heavy machines used for harvesting and transport of trees. The tires should also have improved floating aility on soft ground and snow.

On account of their mode of operation, the pulling wheels for transport by means of skidders should have a design and size different than the wheels of vehicles with carrying functions e.g. heavy forwarders or mobile processing machines with a total weight of 20 tonnes or more.

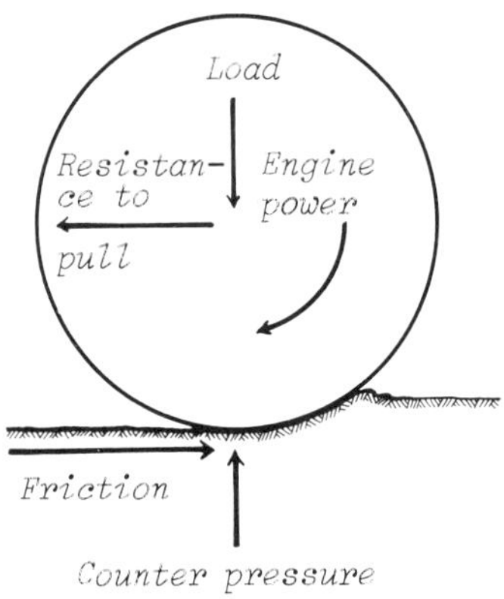

FIGURE 167. Forces acting around the wheel. A large part of the available motor power is consumed for the compensation of the resistance to rolling (friction) which increases when wheel is sinking into soft ground and when diameter of wheel is small (Hedegård, 1951).

Obstacles

Obstacles of various kinds often encumber travel in terrain. Some obstacles can be passed if the vehicle has a high clearance, approximately 35 cm or more.

Other obstacles may be passed over without the vehicle overturning. This requires good stability which depends on width of the vehicle and height to the centre of gravity in the load. In rough terrain, the forest tractor must be rather wide, often more than 2.50 m, while a farm tractor is approximately 1.60 m wide.

A third way of clearing an obstacle is to drive around it. Maneuverability around an obstacle depends on i.a. the capability and principle of steering and the turning radius.

Slopes in terrain

Dependability of the tractor in slopes is an important feature of the capability to travel in terrain. Driving in side slopes, which should be avoided, is dangerous because of the risk of rolling over. This applies in particular when speed is higher than normal which gives a strong momentum with direct adverse effect on stability (Trzesniowski, 1981).

Other kinds of slopes can be down-hill or up-hill. Driving in up-hill slope is often dangerous because of the risk of bucking.

The ability of the tractor to clear down-hill and up-hill slopes has been improved greatly in recent years by means of four-wheel drive and good traction due to bigger loads. This development partly turns upside down the old concepts concerning the outermost branches of the road system with respect to allowable gradients.

Carrying capacity

At transports on soft surfaces the specific ground pressure (N per cm^2) should be as low as possible. The desirable low ground pressure can be achieved by using wide tires, or several wheels in a suitable combination with (or without) track equipment.

For a tire size of 16.00 - 24, the numbers give the nominal width of the tire and diameter of the wheel rim (in inches), the carrying surface is approximately 0.46 m^2. This gives a ground pressure of 11 N per cm^2 at an axle pressure of 10 tonnes on the rear carrier, corresponding to the weight of 150 m^3 of piled pulpwood in the loading space. Wider tires will reduce the ground pressure to 5 N per cm^2 which is a pressure that can be normally sustained by soft ground. Carrying capacity of the ground can be improved by putting limbs in the paths of transport (Scholander, 1972).

High hauling capability required

The hauling capability of a tractor depends not only on the weight and engine power of the vehicle but also on the type of road surface.

The theoretical hauling capability of a vehicle can be calculated if the crank axle output (kW) of the engine is known. The drive shaft output can be obtained by reducing the crank axle output by approximately 10 percent for transmission losses in gears.

Torque of the pulling wheels depends on the number of revolutions of the axle per minute (n). The hauling power (P) is obtained according to the following general formula if radius of the pulling wheel (r) in metre is known:

$$P = \frac{kW \times k}{n \times r}$$

where k converts kW into Nm per revolution and minute. Expressed differently, the theoretical hauling capability is the ratio of the converted value of engine torque and the radius of the rolling wheel.

Practical hauling capability

The practical hauling capability is obtained by multiplying the traction coefficient by the ground pressure of the pulling axle. It is possible in several ways to improve the traction of the hauling vehicle by special measures such as sharp cleat arrangements. The ground pressure of the pulling axle can be improved by means of loading cleats.

It is important in a given transport situation to know the theoretical hauling capability in order not to require more practical hauling capability than that which can be sustained theoretically. What would happen otherwise?

When the hauling capability of a tractor vehicle is high, it follows that a high loading capacity can be achieved. However, this requires a load carrying vehicle adapted to the assortment and to the quality and distance of the roads. An optimum load with a relatively low centre of gravity gives good stability.

Cranes and winches

The means available for loading, unloading and other handling of timber will be discussed here with primary attention to cranes and winches.

Cranes

There are two types of cranes available for loading and unloading of timber on forwarders and skidders with clam bunk viz. knuckle boom cranes and telescopic cranes, both types operated hydraulically. The combination of knuckle boom crane with telescopic cranes is also used to some extent.

Knuckle boom cranes

The knuckle boom cranes are predominant in Europe. This type of cranes was introduced for practical use in forest operations in the beginning of the 1960's. Cranes equipped with a hydraulic grapple can be used for quick and efficient loading and unloading. The terminal time has been reduced essentially and the tractor vehicle has become competitive, striking out the horses from normal terrain transports of timber (Krivec, 1972).

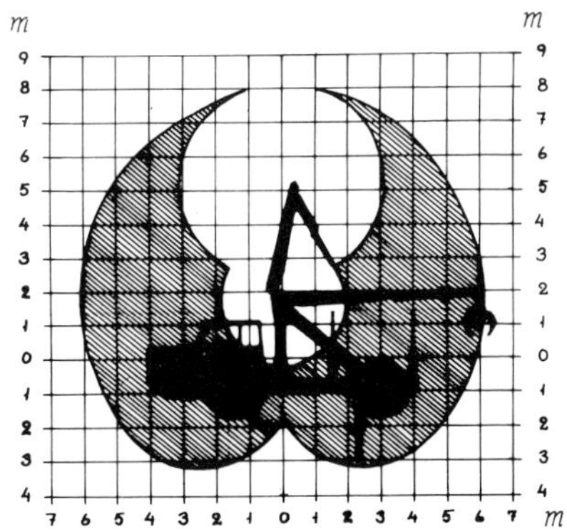

FIGURE 168. The reach or working area of a crane can be demonstrated by means of a sweep graph.

Characteristics of the crane

Reach of the crane, lifting capability, grapple area, and maneuvering speed are all features of great interest (Figure 168).

Hydraulic knuckle boom cranes and grapples have been developed rapidly since the end of the 1950's into some of the most important machine components in mechanized tree harvesting operations. Modern cranes used in tree harvesting are not likely to be replaced by any other tool or machine in the foreseeable future.

A good crane in tree harvesting is expected to provide a correct geometry of movements and correct reach for each particular work situation and also a large capacity, dependable functioning and long durability. Below is given an example of a small flexible knuckle boom with great capacity. It is suitable for small forwarders or clam bunk skidders.

Technical data:		
	Lifting capability, gross:	60 KNm
	Turning capacility, net:	14.7 KN
	Recommended capacity of pump:	1 l/s
	Max. pressure in operation:	19.6 MPa
	Weight, knuckle boom with 0.35 grip:	1460 kg

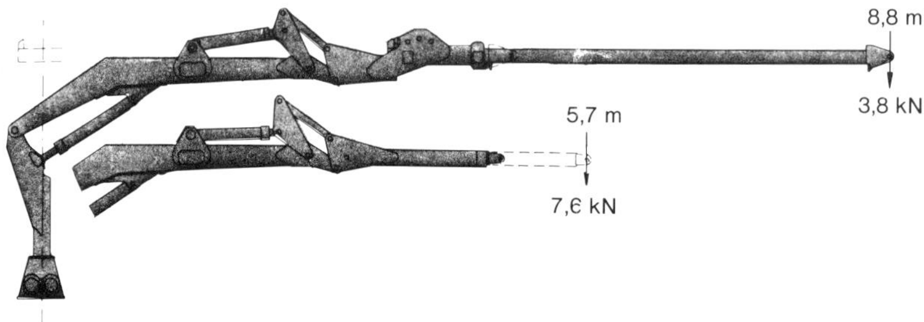

FIGURE 169. Separate picture of knuckle boom or knuckle arm crane.

FIGURE 170. Crane mounted on harvester.

A crane boom is usually made of cold-seasoned special steel and designed with a box profile. The steel provides a very rigid construction with respect to bending and turning.

A turning mechanism with double cylinders and 'floating' pistons provides a dependable operation. The whole head is protected in its movements by an end damper.

A crane tilt design retains the turning power at inclinations up to 36 percent, which gives improved flexibility, less risk of damaging the remaining trees and less equipment wear.

Work with this crane is easy on the operator and the machine if it can be done by gentle and accurate movements e.g. by means of double lever control with proportional steering and individual adjustment of all functions under constant pressure in the hydraulic system (ÖSA System, 1981).

The grapple area or the maximum space between the grapple shanks determines i.a. the size of the timber bunch and, hence, it is of interest for the bunching and collection of timber.

The maneuvering speed of crane movements such as positioning, lifting movements, knuckle boom movements, extension boom movements, grapple turns and grapple movements all influence the cycle time per bunch and, hence, the total time of loading and unloading.

Steering levers

Modern steering levers in machines used for tree harvesting usually have a grip designed on the basis of ergonomic studies. This grip provides an easy and relaxed working position. Shoulders, neck and arms are not excessively strained. The grip may be molded in plastic material and it gives a quick adjustment to the temperature of the hand. Use of two levers provides an electro-hydraulic control of six functions by proportional steering.

Light lever handling and controlled lever positions give easy and convenient work. Light weight and simple cable connections facilitate adjustments of the lever positions to various desirable situations and steering will be precise and quick in the various mechanisms (Figure 171).

FIGURE 171. Steering lever designed on the basis of ergonomic studies.

The lifting capability or lifting moment is the product of lifting power and length of crane boom.

Moment = power x length of boom

The upper limit of the lifting capability is determined by the strength of the crane boom and its base and by the pressure in the hydraulic cylinder of the crane boom.

The lifting capability, which determines the size of timber bunch, is often expressed in terms of tonne-metres. A crane with a lifting capability of 5 tonne-metres can lift 5 tonnes with a boom length of 1 m, or 1 tonne with a boom length of 5 metres from the pivot of the crane.

Winches

The winch is an important piece of equipment for e.g. bunching of trunks to a skidder. The winch is also used on forwarders e.g. for loading of bucked timber. Winches are necessary for towing aid to other machines.

The capability of a winch is determined by two factors viz. torque of the drum and speed of the cable (Cf. Wassilev, 1981).

Torque depends on diameter of the drum and it is influenced by the number of layers of cable on the drum. Speed of the cable depends on the diameter of the drum and the number of revolutions per time unit (rpm). (Figure 172).

Loading by means of winch can be done by remote control via electric wire or by radio. This arrangement was common occurrence on winch cranes mounted on tractors already in the beginning of the 1960's. Loading can be done by one man only.

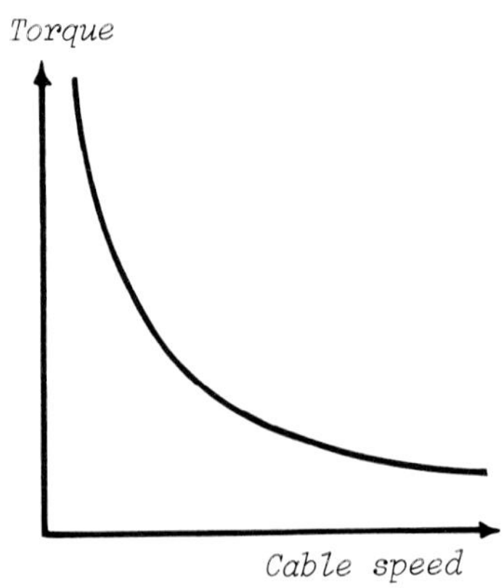

FIGURE 172. Principle relationship between torque and cable speed in a winch.

Methods of transport in terrain

Skidders and forwarders

A method of transport is chosen on the basis of conditions defined by the objects of transport, routes of transport in a given environment and means of transport. The method is finally designed with attention to economic considerations.

Methods of transport applicable to forwarding, primarily in terrain, can be classified with respect to the character of the transport objects. It is thus a matter of transporting whole trees, tree length trunks, assortments (short timber) and chips.

Horses (animals), simple tractors and forwarders are mostly used for transports of assortments (short timber) in terrain. The proportion of these means of transport varies largely from one country or region to another depending on methods of felling used and on the general state of transport technology.

A country in the northern coniferous region where the assortment method dominates the forest operations has reported the following proportion of various means of transport in terrain in 1971 and 1981:

	% in 1971	% in 1981
Horses	5	3
Simple tractors	25	17
Forwarders	70	80

The total number of transport units used in large scale forest operations in the same country in 1969, 1971 and 1981 is shown in table 10.

Current trends concerning methods are reflected in the changes of machine equipment from 1969 to 1981. The same trends generally apply to the small forest owners.

Methods of transport by means of tractors

Primarily two types of tractors can be distinguished for transports in terrain.

Table 10. The total number of transport units used in large scale forest operations in 1969, 1971 and 1981.

Type of machine	No. units		
	1969	1971	1981
Tractors equipped with grapple loaders forwarders		2 900	4 600
Farm tractors		700	300
Total	3 200	3 600	4 900
Simple tractors	700	300	100
Clam bank skidders		40	160
Other skidders	550	560	100
Total	1 250	900	360
Horses (approximately)	1 200	400	100

The reason why the harvesters are increasingly replacing the combination feller-processor (delimber and bucker) are:

1. The harvester has the functions of a base machine at a lower investment cost.
2. The harvester requires only one operator which means lower cost of operation.
3. The harvester can operate continuously which means that time is saved.
4. Trees felled by means of the harvester can be placed immediately into position for infeed. The trees are not laid in piles on the ground where they are frozen stuck in winter, which happens when the felling is done one or several days before further processing.
5. Tree harvesting is simplified.
6. A harvester is no more complicated than the combination feller-processor; -rather the opposite.

FIGURE 173. Skidder with components of power unit and load carriage.

Transport of trees by means of tractor

Whole trees are transported over relatively short distances. At present this method of transport is applied to approximately one percent of the total volume of timber harvested in Sweden.

Transport of whole trees to strip roads within 100 m distance

Transports of whole trees within a distance of 100 m to strip roads can be carried out by means of e.g. a feller-skidder. Such a machine is equipped with a clam bunk and a felling mechanism mounted on a hydraulic knuckle boom or telescopic boom. After felling the trees are transported to strip roads for direct unloading of the whole bunch which is then processed by means of a machine operating along the strip roads.

Transport of whole trees within a distance of 400 m

Transport of trees to a major landing or a place of processing, if done within a distance of 400 m, can be carried out by means of e.g. a feller-skidder, or at manual felling by means of a skidder equipped with clam bunk or winch.

Skidders equipped with winch

Transports by means of tractors equipped with winch, cable and cable (or chain) chokers in combination with felling are relatively labour intensive operations. In recent years, feller-skidders or skidders equipped with clam bunks have been introduced.

Felling for skidders equipped with winch is carried out so that the trees can be hauled butt end or top end first.

Depending on its size, the tractor may be equipped with a single drum or a double drum winch. If a double drum with each drum operated separately is used, a half load at a time can be hauled at full power on the cable. Hauling power usually varies from 20 000 N - 30 000 N to 60 000 N - 10 000 N. The highest speed of the cable is 60 - 70 m per minute and length of cable is usually 50 m.

Removal of chokers (or chain couplings) at unloading is a rather time consuming procedure.

Skidders equipped with clam bunk

All work with chokers or chain couplings is eliminated when skidders with clam bunk are used. Skidders of this type are preferable for transport of whole trees butt end first over long distances, particularly when trees are big, terrain is easy and ground has a good carrying capacity.

When this type of machines is used, centre of gravity of the load is moved in over the rear axle providing for reduced friction and improved hauling power.

This type of skidder has a clam bunk with flexible braces (jaws) mounted on the rear of the tractor vehicle (Figure 174). Generally, the clam bunk has a wide and low profile with a relatively large grapple area. Loading is done by means of a regular hydraulic crane mounted adjacent to the driver's cabin on skidders with a long wheel base.

Timber is held by the braces of the clam bunk by means of sharp edges on the inside of the braces or by means of chains or cables built into the braces in order to suspend the load of timber to some degree. Loading and unloading is quicker when clam bunks are used than when winches are used.

FIGURE 174. a) Skidder with clam bunk.

FIGURE 174. b) A heavy clam bunk skidder with a load capaity of 18 tonnes and a hauling capability of 22 tonnes or 25 tonnes. It has 8 wheels with tracks over the front and rear wheels for operations on rocky ground or on sites with a low carrying capacity.

FIGURE 175. Skidder with clam bunk used in the tree system.

Skidders equipped with grapple

A skidder with grapple is another type of tractor designed for transports of trees or trunks. The grapple used on this type of skidder is of a design different than that of the regular grapples of skidders or forwarders. This grapple is designed for collection of scattered trees, trunks or bunches of timber primarily by reciprocating motions along the vehicle. Skidders with grapple are generally quicker in loading and unloading than skidders with chokers.

Transport of tree length trunks

To some extent transport of tree length trunks is carried out by means of skidders with winch. Usually, hauling is done top ends first. The friction of butt ends is rather great and a high practical hauling capability is required (Table 11).

Table 11. Average friction coefficients at skidding of trees and tree length trunks with lifted points of pull (Björklund, 1968).

Object of skidding	Point of pull	Ground surface		
		Bare ground with vegetation dry	moist	Snow 25 cm
Trunks				
Pine and spruce	Top and butt	0.70	0.50	0.40
Trees				
Pine	Top	0.75	0.60	0.45
Spruce	Top	0.80	0.65	0.50
Pine	Butt	0.80	0.70	0.55
Spruce	Butt	0.85	0.75	0.60

Skidding by means of winch

Transport of trunks is to some extent done by means of skidders equipped with winch. Similar to other forest tractors the skidders with winch are being equipped with bigger wheels. This trend would mean less theoretical hauling capability which is the ratio of the torque on the pulling axle and the radius of the wheels. However, engine power has increased relatively more than wheel size. There are now skidders with a practical hauling capability of 150 000 N with heavy load and a high traction coefficient.

For the purpose of loading trunks, the trees are usually felled so that the top ends are collected for quick choking or coupling with chains.

Transport of trunks can be carried out directly from the stumps or from strip roads to landings, to a place of processing or reloading to trucks for transport to a central processing establishment.

Skidding by means of winch has previously been a common method of transport when felling, delimbing and topping has been done manually.

Skidding by means of clam bunk

The transport alternative with skidding by means of clam bunk, butt ends first, is a good solution for transports from the strip roads when delimbing has been mechanized and the trunks have been pulled evenly into bunches with butt ends in the direction of transport. Due to the relatively short time required for loading and unloading and the high load capacity, this method of transport can be used for longer skidding distances than those possible when winch is used. This method of transport, therefore, is gaining application.

Transport of assortments or timber bucked into multiple length

Transport of assortments is a common method of timber transports in terrain, but in some cases transport of trunks by means of forwarders i.e. tractors carrying the whole load above the ground, is more advantageous. Two main types of forwarders are distinguished viz. wheel forwarders and track forwarders.

Wheel forwarders

Engine power of wheel forwarders varies from 26 kW to 118 kW (Figure 176). The maximum load capacity ranges from 9 tonnes to 15 tonnes when the forwarder is operating in terrain, equivalent to 15 m^3 - 25 m^3 of pulpwood. The weight of loads at transports on road ranges from 10 tonnes to 20 tonnes, equivalent to 16 m^3 - 32 m^3 of pulpwood. Ground pressure of wheel forwarders at full load is 6.6 N - 10.5 N per cm^2 for the front vehicle and 10.5 N - 18.0 N per cm^2 for the rear vehicle with two wheels, for bogie tracks 5.0 N - 6.0 N per cm^2.

Track forwarders

Engine power of track forwarders varies between 26 kW and 81 kW. These machines have half, three-quarter or full tracks running over the rubber tires of the front vehicle. Occasionally, tracks are also used on the bogie of the rear vehicle.

FIGURE 176. Hemek 650, a forwarder with 8-wheel operation, four wheels on the tractor and four wheels on the carrier. The carrier is powered from the pto (power take-off) of the tractor via drive rollers placed between the bogie wheels.

The maximum load weight of track forwarders varies between 6 tonnes and 10 tonnes, equivalent to approximately 10 m^3 - 16 m^3 of recently felled timber.

Due to the tracks, the ground pressure of the track forwarders is lower than that of the wheel forwarders. Pressure at full load varies for the front vehicle between 3.0 N per cm^2 and 4.0 N per cm^2 and for the rear vehicle between 5.5 N and 18.0 N per cm^2.

On the basis of a comparison with the ground pressure of a skidder, it may be stated that the ground pressure of the loaded rear wheels of a skidder is approximately equal to the ground pressure of the wheel forwarder. Moreover, skidding is causing momentarily high ground pressure values, which may cause bogging down on ground with a low carrying capacity.

A 16-wheel forwarder for difficult terrain

The forwarder has good travel ability on peat bogs, in deep snow and in steep and rocky slopes. It is 25 percent - 30 percent faster than conventional forwarders. The wheels have been equipped with specially designed rubber tracks. The machine was developed in 1983 for difficult transport conditions.

FIGURE 177. A 16-wheel forwarder for difficult terrain.

High load capacity

The weight of loads mentioned above indicates that a relatively large amount of timber can be transported by tractors on strip roads to landings at the truck road.

High load capacity requires efficient bunching and concentration of timber along the strip roads and it demands high standards with respect to width, straightness, slopes, eveness, carrying capacity etc. of the road.

Transport performance

Transport performance is usually expressed in terms of volume per unit of time e.g. m^3 per hour for a given distance of road. Output of transport can also be presented by e.g. the expression tonne-km per hour.

Calculation of the transport performance

A calculation of the transport performance is based on the product of the number of loads transported per time unit and the load size. Every detail of the actual transport situation must be analyzed if it is desireable to influence time per turn and size of load in order to conscientiously improve the transport performance.

Since there are no two transport situations that are exactly identical, planning and organization of all transport must be treated separately for each case. For this reason, it is necessary for the transport operator to know certain general relationships with respect to transport techniques.

Transport factors

If the factors, which are presented in the introduction to the chapter on transport of timber (Figure 140), are used as a base, it is found that transport is primarily influenced by three main groups of factors:

Technical factors influence the output in m^3 per hour. This group includes such factors as travelling (driving) time, terminal time and size of loads

Organizational factors influence the degree of utilization (hours per year). This group includes factors such as planning, training of operators, procurement of transport equipment, and service.

Economic factors influence directly the transport costs. This group includes factors such as costs of machines and personnel, collection and processing of performance data and comparative analyses of machine costs.

The influence of these factors on the costs of transports has been defined in a formula. Efforts should be directed in order to achieve as low values as possible in the numerator and as high values as possible for the factors in the denominator in order to obtain the lowest possible cost. Transport is also influenced by biological and environmental factors.

What primary questions should be raised for a close study of the factors concerning time per turn and load size which are of great importance for transport performance?

Technical factors of transport

Time per transport turn consists of travel (driving) time and terminal time.

Travel time

Travel time consists of travel time with load and travel time idle. These components of time e.g. expressed in minutes, are actually equivalent to a ratio of the factors road distance in metres and speed of travel in metres per minute.

Thus:

$$\text{Travel time, min} = \frac{\text{road distance, m}}{\text{speed of travel, m per min.}}$$

To reduce the travel time, the transport operator has to reduce the road distance and/or increase the speed of travel.

Road distances

Road distance in terrain can be reduced by developing the system of truck roads according to the analysis presented in a previous chapter for determination of the optimum density road system in a given region. Road distance in terrain is also reduced by good planning of the strip roads, aiming at a feasible pattern of roads for a given topograpy by means of straightest possible strip roads perpendicularly to the truck roads. Density of the road systems must be geared to i.a. the costs of road construction and to the travel costs.

Speed of travel

Speed of travel is influenced by design and route of the road and by the capability of the tractor-vehicle to travel in terrain.

Among road features required for increased travel speed may be mentioned a smooth and almost horizontal road surface. Side slopes and steep uphill or downhill slopes should be avoided. A sufficiently wide road surface and good carrying capacity are other requirements. The carrying capacity of the strip roads can be improved by means of limbs.

Among tractor features required for good operation in terrain may be mentioned good steering and short turning radius (4-7 m). The vehicle should also have good carrying capacity i.e. sufficiently large contact surface between wheels (tires) and ground. Sharp edges, protruding parts or auxiliary equipment that extends outside the load, getting stuck in remaining trees, must be eliminated. The vehicle should be 'stream-lined'.

Traction

Speed of travel is largely influenced by traction. Poor traction which results in spinning wheels can reduce the speed to stalling.

Traction can be influenced directly by means of certain tire tread patterns or by means of anti-slip devices (Figure 178).

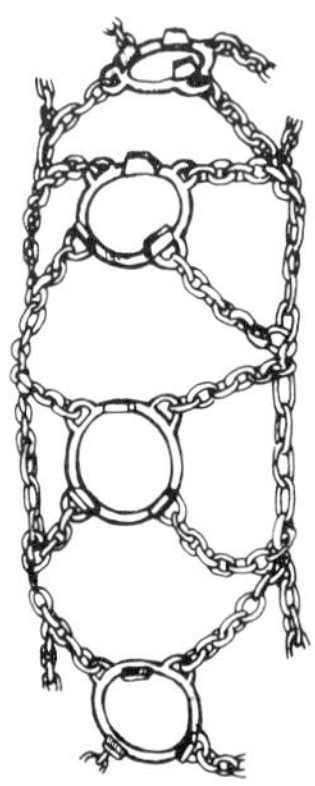

FIGURE 178. Anti-slip designs of chains and cleats (left and below).

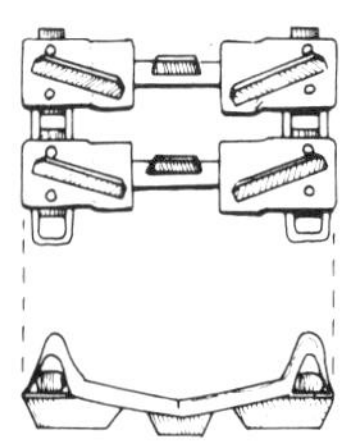

FIGURE 179. Track plates with cleats for operation on soft ground or snow.

Anti-slip devices mounted on several wheels improve traction further (Figure 179).

A correct distribution of load on the pulling wheels and tracks also improves operation.

Traction can be influenced indirectly by preparation of the road surface e.g. by means of limbs - a simple and inexpensive measure -, gravel, sand or other material that can give improved traction in certain critical locations.

Terminal time

Terminal time included in the time per turn consists of time for loading (also when moving) and unloading.

Terminal time for loading

Time required for loading depends on several factors that can be influenced by the transport operator. Loading time consists of indirect time, positioning, and direct time.

The total time required for loading a truck is the product of cycle time per bunch of timber that e.g. the grapple loader puts on the truck and the number of bunches necessary to fill the load. This is also transport although on a scale smaller than that of travel.

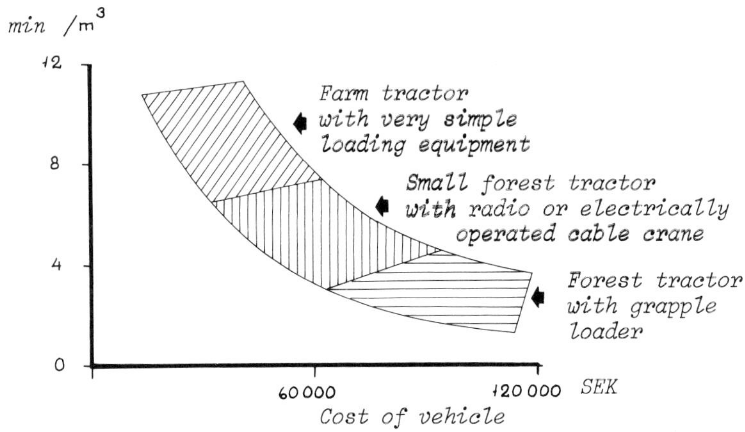

FIGURE 180. Principal influence of equipment on time of loading assortments.

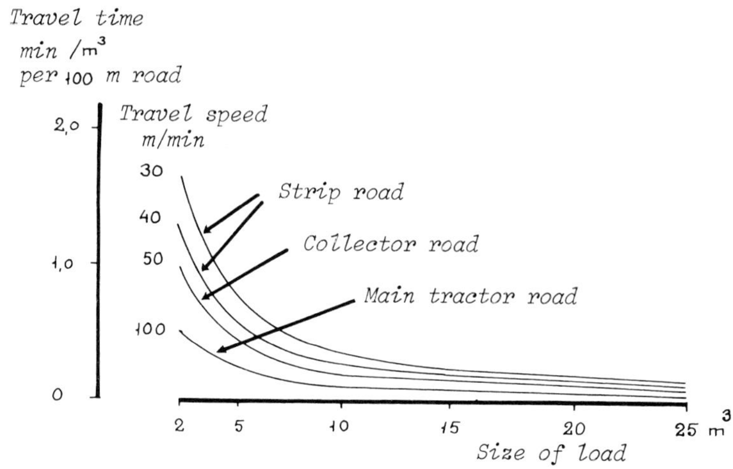

FIGURE 181. Influence of road class and size of load on travel time.

Loading of assortments (short wood) or bucked timber

Time required for loading of a bunch or a log also consists of travel time and terminal time. Travel time is determined by the average distance between centre of gravity in the load and the places where piles or bunches of timber are located, and by the operational speed of the knuckle boom or telescopic boom along the various curved movements.

Terminal time per bunch includes time for loading of grapple and unloading onto the carrier. Loading time is reduced by means of quick grapples and if the bunches and piles of timber are easily accessible (Figure 180).

Size of the bunch in each lift depends on grapple area, lifting power, capability of the crane, and by the size and accessibility of the timber piles.

Loading of trees and tree length trunks

Loading of trees and tree length trunks can be carried out according to the same principles concerning various partial movements as those applied for assortments or bucked timber. Conditions that influence the volume of loads at transports of trees and trunks are shown in Figure 182.

The upper part of the dashed region in Figure 182 represents favourable transport conditions while the lower part of the region represents poor transport conditions. When whole trees are transported by means of the same hauling capability as that used for transports of trunks, load capacity is reduced by 20 percent primarily because of increased friction, bulkiness and amount of limbs.

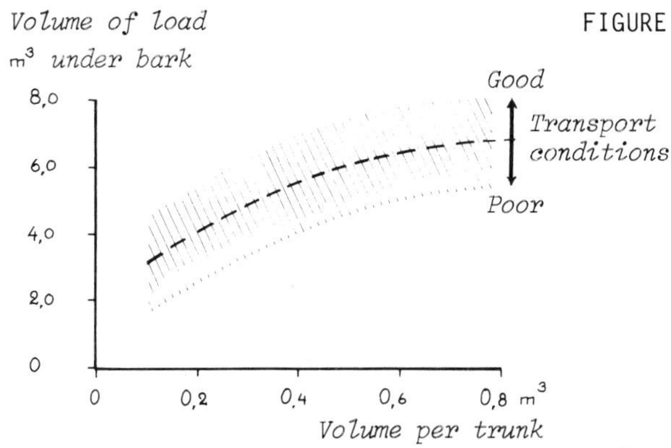

FIGURE 182. For a tractor with given hauling capability, volume of load is influenced by transport conditions such as terrain, slopes, amount of snow, timber quantity and size of the trunks.

Terminal time for unloading

Terminal time required for unloading is influenced by the layout of the landing and by the method of unloading.

Layout of landing

Size of the landing should be adapted to the expected flow of timber and buffer storage. The landing should be well graded and have good carrying capacity.

Method of unloading

The method of unloading is chosen on the basis of:

- Load carrying equipment and loading arrangements
- Requirements concerning drying of timber and piling
- Method of further transport, eg. truck, flatbed or trailers
- Availability of separate equipment for loading onto trucks, or trucks equipped with grapple loaders.

Time required for unloading is influenced by the design of ...ces carrying equipment with respect to bunks, stakes, loading sp... height, separating dividers for different assortme...

etc. Unloading time is also influenced by the position of the crane on the vehicle in relation to the load carrier as well as the crane operator's overview and control of his work. All these conditions are details that require analyses aimed at reducing terminal time.

Time required for unloading by means of grapple is sensitive to size of timber, time per m^3 declining when the average diameter of the timber increases.

Time required for tipping is rather insensitive to size of timber, time per m^3 being largely equal for small and big timber.

At rush unloading half the load is dumped while the rest must be unloaded by means of grapple, or manually. Time required for unloading by this method is influenced by the size of the timber to an extent intermediate to the time required for the two methods described above.

When trunks and trees are transported by means of skidders equipped with clam bunks, time required for unloading is relatively insensitive to size of timber. As is the case in tipping, unloading is very quick.

When trunks and trees are transported by means of skidders equipped with winch, time required for unloading is very sensitive to the average size of timber.

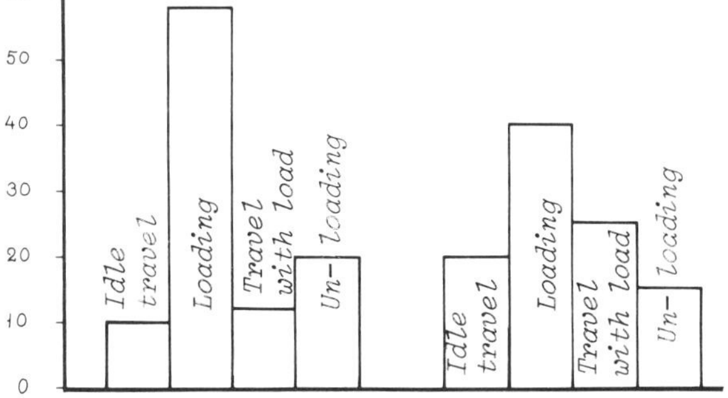

FIGURE 183. Relationship between travel time and terminal time for different distances of transport in terrain.

Relationship between travel time and terminal time

The relative proportions of travel time and terminal time for different -es of transport are shown in Figure 183.

Time required for idle travel and for travel with load also depends on the speed of travel over the distances of transport. The comparison given in the graph is based on equal speed of travel. Time required for loading and unloading depends on the method used and on the size of load. Figure 183 may be interpreted further.

Size of load

Conditions that influence the number of turns per unit of time has now been discussed. However, it is also of interest to study the amount of timber that is transported in terrain for each turn, i.e. load size, which is the second factor of great importance for transport performance. Size of load depends on two factors:

- the practical hauling capability
- the optimum load capacity

The practical hauling capability depends on the traction coefficient and ground pressure underneath the pulling wheels or tracks.

Traction

Depending on the moisture conditions in the ground, the traction coefficient for bare ground varies between 0.6 and 0.8 on hard surfaces and between 0.2 and 0.5 for forest ground when tires are used on the pulling wheels.

In winter traction coefficient varies between 0.3 and 0.4 for firm and sanded snow roads and betweeen 0,1 and 0.2 for soft snow roads and ice roads when tires are used on the pulling wheels.

If anti-slip devices are mounted on the tires, traction coefficient is raised by 10 percent, and if tracks are used, another 10 percent.

Ground pressure

Ground pressure i.e. pressure exerted on the contact surface between the pulling wheels or tracks and the ground primarily depends on the vehicle load. The vehicle load is here meant to be the total weight of the vehicle itself and the load above the pulling wheels or tracks (Figure 184).

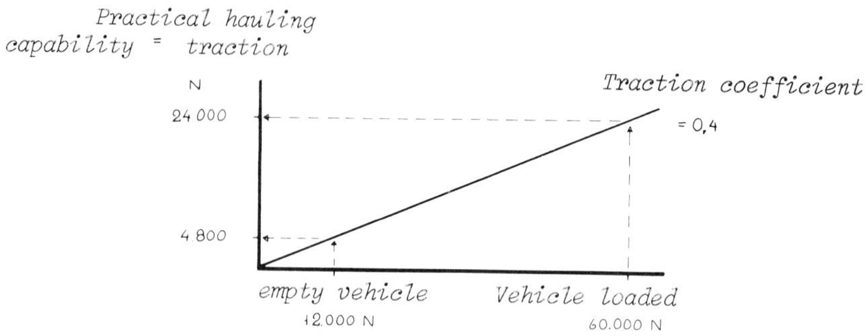

FIGURE 184. Relationship between the practical hauling capability and weight of vehicle with load.

Practical hauling capability

The practical, available hauling capability of a machine has an upper limit i.e. the theoretically available hauling capability is not to be exceeded.

Optimum load capacity

Studying the specifications of the vehicles available for transport of timber, the planner must establish the hauling capability of the vehicles. Moreover, he must know the forces resisting the movements of machines in transport operations. To some extent resistance to movements influences the optimum load capacity.

Various types of resistance to movements that may be considered at planning of transports are:

- Slope resistance, positive and negative
- Resistance to skidding
- Resistance to rolling
- Air resistance at high velocities and high winds
- Acceleration resistance (inertia)

Objective of transport operator or planner must be to minimize all losses of motor power, i.a. by measures designed to reduce various influences of resistance.

Slope resistance

The slope resistance depends on the gravitational pull and varies with the weight of the vehicle and the gradient of the slope. Slope resistance can be calculated as a product of slope coefficient and weight of vehicle.

If inclination is 1:5, slope coefficient is 0.20. If weight of vehicle with load is e.g. 100 000 N, an additional hauling capability of 20 000 N is necessary to compensate for this uphill slope. An opposite force of the same magnitude is influencing the movement downhill.

Resistance to skidding

Resistance to skidding has to be overcome primarily at transports by means of carriers with runners, and at skidding of timber.

Resistance to rolling

Resistance to rolling occurs for vehicles with wheels or tracks because of deformation of the pulling wheels or tracks such as compression of tires, deformation of the ground leading to rutting, and because of friction in anti-slip devices and tracks.

Resistance to rolling is the product of load and a coefficient of resistance to rolling. This coefficient for forest ground amounts to approximately 0.06 - 0.10. Resistance to rolling (coefficient) increases when the wheels or the tracks sink into soft ground. It decreases when the wheel size increases, bigger wheels providing for better carrying capacity and "absorbing" minor obstacles on the ground.

Total and maximum resistance to movements

Total resistance is obtained by summation of slope resistance, resistance to skidding and resistance to rolling. The maximum resistance can then be compared with the theoretical or available, hauling capability. This can be done by adapting the load size in order to make sure that a certain margin of hauling capability is available. This is necessary in order to counteract the dynamic forces manifesting themselves at transports of heavy loads.

Thus, load size depends on method of transport i.e. if load is totally carried on wheels or runners or if load is totally or partly skidded. Resistance to skidding also varies with degrees of processing at delimbing and debarking and for varying length of timber and units of handling.

Organizational factors of transport

The organizational factors of transport primarily influence the degree of vehicle utilization. A maximum use of available machines is required for good transport economy. Utilization can be expressed in e.g. hours of production per year.

At high vehicle utilization it is primarily the proportion of capital costs in the total costs that declines. A high degree of vehicle utilization can be achieved by good planning and organization, well trained personnel, good machines and tools, good maintenance and efficient service.

Planning and organization

Planning aims at a concentration of transports to consolidated harvest areas i.a. in order to reduce inefficient and expensive transports over long and many roads between several scattered areas. Within the harvest area a division into partial, alternative areas of transport can be made for operations in summer, autumn, winter and spring, taking into account the carrying capacity of the ground in order to achieve a year-round transport program.

Well trained personnel

A proposed transport plan cannot be implemented unless all personnel involved is adequately trained.

Good machines and tools

To attain a high degree of machine utilization, vehicles and tools must be well adapted and designed, durable and dependable. The final choice of a transport machine is an important decision. Matters concerning spare parts, maintenance and repair service must be decided with a view to reducing stoppage and breakdown time.

Some desirable ergonomic and technical data on a modern forwarder

1. Low noise level in the operator's cabin, if possible down to 76 dB(A). A low noise level can be achieved by a rubber suspension of the engine

2. Safety glass strong enough to replace a protective screen in order to provide free field of vision
3. Easy maneuvering panel accessible for quick service
4. Sufficient lighting by correct mounting of e.g. 10 rectangular halogen lights for night time operation
5. Automatic sprinkler system

Technical data for two different forwarders

Engine output:	55 kW	62 kW
Maximum load:	7.0 tonnes	7.5 tonnes
Driving speed on access roads:	20 km/h	30 km/h
Crane pump capacity:	45 l/min	70 l/min
Clearance, front:	485 mm	560 mm
Noise level in cabin:	82 dB(A)	76 dB(A)
Price excl. tax, 1983:	526 300 SEK	598 000 SEK
(1 $U.S. = 8 SEK)		

Economic matters

To attain the objective of lowest possible transport costs according to formula, correct costing of the transport work is required.

Performance data

Costing of transport requires i.a. a number of reliable data on the performance of various machine types. The performance data are usually obtained from some organized records taken at ergonometric studies or from technical analyses of machine designs.

Costs of capital and operation

The costs of capital and operation of the machines must also be known, including information on the direct and indirect wages of the operators.

Finally, knowledge of comparative analyses of various transport alternatives is required in order to arrive at the best system.

Relationship between terminal costs and travel costs

When analyses of transport have disclosed the terminal costs and the travel costs for various alternatives, it will be obvious that the relationship between these costs varies for different alternatives (Figure 185). Transport alternative I with low terminal costs e.g. due to quick loading and unloading, has high travel costs because of low load capacity. Alternative II shows the opposite.

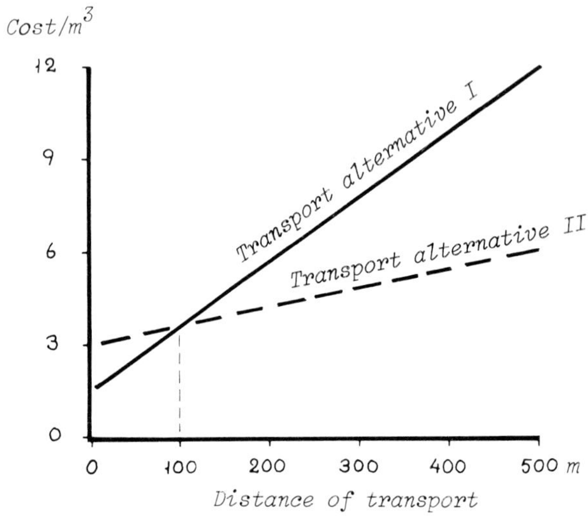

FIGURE 185. Relationships between distance of transport in terrain and costs of two alternatives. Alternative I has low, alternative II has high terminal costs. Alternative I has high, alternative II has low travel costs.

Trends of transports in terrain

The following presentation is a brief description of recent development and performance of transports in terrain by means of tractors.

Development of a forest tractor

In 1960 vehicles hauled by horses were still a dominant means of transport in the forests (in Sweden). Vehicles on wheels had been developed for bare ground operations.

Meanwhile, the first advanced courses on bare ground transports by means of tractor were arranged for administrative personnel. The courses aimed at a broad approach to teaching techniques and methods of tractor operation on the basis of tractors and loading equipment available at that time.

Skidders were imported from USA. They were equipped with frame steering and four-wheel drive (big wheels). These machines were subject to time studies of i.a. skidding of trunks (Staaf, 1983).

Subsequently a skidder was combined with a big-wheeled tractor trailer built for transports of bucked timber (Figure 186).

FIGURE 186. A forest tractor with big-wheeled trailer, telescopic crane and cable grapple. A prototype forwarder.

In comparisons with the modified farm tractors the new vehicle combination showed superior travel capability with big loads. However, the machine with its six big wheels appeared to have a superfluous pair of wheels.

The first forest tractor

In Sweden the first forest tractor was built in 1962. A trailer with big wheels was coupled to a farm tractor, the front wheels of which had been removed (Figure 187).

The machine was equipped for frame steering by means of two hydraulic cylinders, and a hydraulic knuckle boom with grapple for retrieval of timber around the machine. The machine was put into manufacture and quickly distributed. A rapid development of forest tractors started among other machine manufacturers (Figure 188).

FIGURE 187. A forest tractor, probably the first one in the world with hydrostatic-mechanic transmission, Filipstad, 1962 (Staaf, 1962b and 1983).

FIGURE 188. Hydrostatic-mechanic components for power transmission in the first Brunett. Filipstad, 1962.

The tractors, both forwarders and skidders, soon increased in size and weight. Wheel sizes were also increased for improved carrying capacity.

The technical evolution has lead to improved performance.

Hydrostatic-mechanic power transmission

One of the most important functions of a modern machine for tree harvesting is the transmission of power from the engine to the mechanisms where it is to be used for specific types of work. The transmission is of decisive importance for the characteristics and performance of the machine. The power transmission of the hydrostatic gear box is a combination of two well tested methods. Power is transmitted hydraulically from the engine to a distribution gear box from which it is further transmitted mechanically e.g. to the pulling wheels. Through the pto the engine can power two hydraulic pumps. One pump delivers the flow of oil for the working hydraulics e.g. to the grapple loader, the hydraulic steering, winch etc., while the other pump is used for the movements of the base machine. By means of a variable displacement, which means that the flow of oil from the pump can be changed, the flow of oil can also be controlled from none to maximum without changing the rpm of the main power source e.g. the diesel engine. However, in this situation the travelling speed of the machine is controlled from zero to maximum. In addition, the direction of the oil flow can be changed which means that the direction of the machine operation is changed from forward to backward. Both the flow of oil and its direction can be controlled easily for various functions by means of a maneuvering stick in the operator's cabin.

Efficiency of the transmission can be measured in percent giving the proportion of machine power that can be utilized by the pulling wheels. Efficiency of the modern hydrostatic-mechanic transmissions can be as high as 80 percent at low gear and 85 percent at high gear when only one wheel axle is engaged (ÖSA).

Due to its high efficiency the hydrostatic-mechanic power transmission provides for a relatively low fuel consumption. The control system also enables two different types of driving viz. constant rpm and controlled speed by the driving stick, and speed controlled by rpm.

Figure 190 shows the design of the hydrostatic transmission.

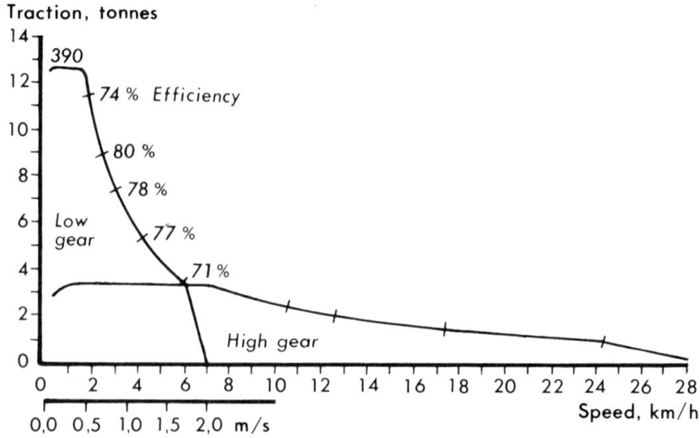

FIGURE 189. Speed and traction graph (actual values at the wheels).

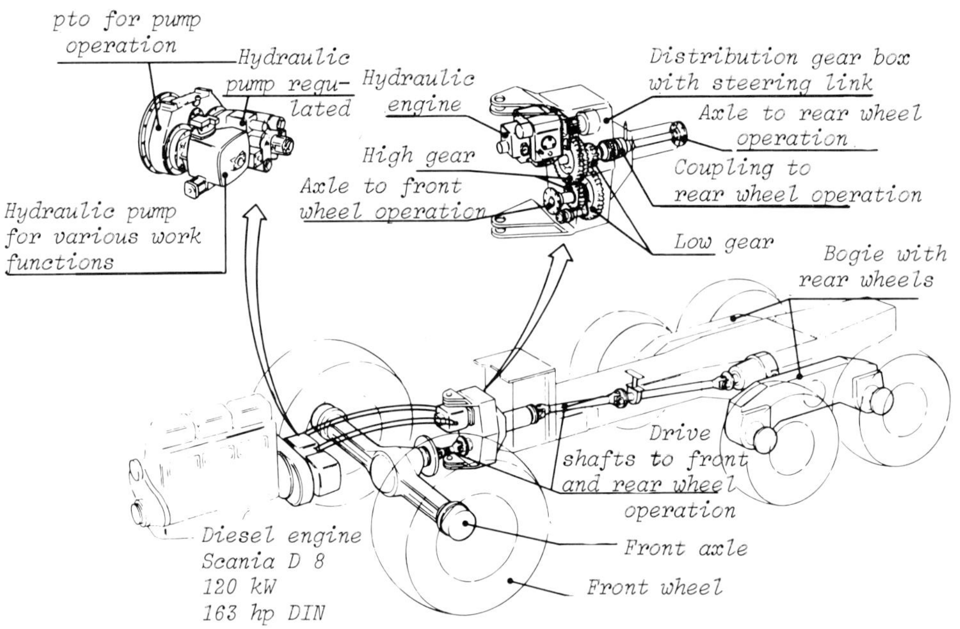

FIGURE 190. Design of the hydrostatic transmission.

Due to the fluid drive of the wheels, travel speed in terrain can be smoothly controlled for each small section of the distance by increasing power on firm ground and decreasing power on short distances over soft ground. This control means an enormously improved travel ability in terrain compared with that possible when a system with mechanic gear box is used which has no synchronization between various gear positions.

Durability of all hydraulic components depends on the work being done. Excessive rpm, excessive heating and impurities must be avoided. Right oil quality is a prerequisite.

Comparison of performance

A comparison of performance in m^3 per hour between and within various types of machines cannot be done fairly. Organization of work places, skill of operators, condition of machines, timber quantity, road distances, terrain difficulties all influence transport performance to a degree varying from case to case.

Further transport in forest operations

Forms of further transport

Further transport of timber from the forest operations to the industries can be programmed in several ways, e.g. for:

- trucks (lorries)
- railways
- river drives
- combinaton trucks + river drives
- combination trucks + railways
- barges or rafts

The proportion of various forms of further transport varies from one country to another depending i.a. on the general development of transport infrastructure, economic situation, distribution of industries, nature of timber resources etc.

While truck transports direct to the industries and railway transports increased in some countries in recent years at the expense of other transports such as river drives, the original form of timber transport, changes in other countries have taken a different direction.

An example of the situation in a north European country in 1970 is given in Figure 191.

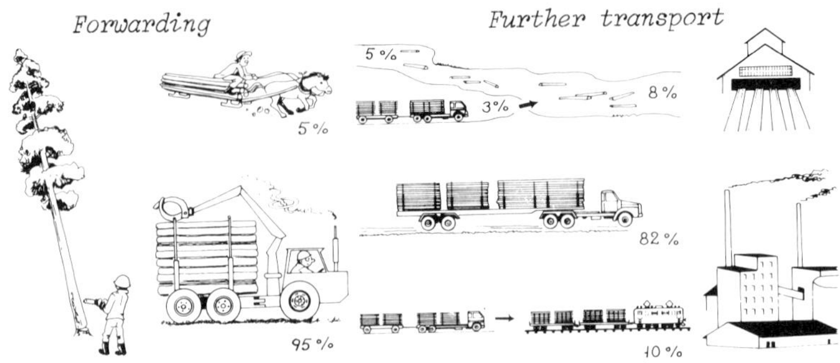

FIGURE 191. Further transport of timber distributed by various forms.

The amount of timber transported to the industries by railwys was then 5.0 million tonnes while transport by river drives was only 4.0 tonnes. The amount of timber transported by trucks was approximately 40 million tonnes. Another 5.2 million m^3 of timber was rafted to the industries or other destinations and only 36 000 m^3 were transported by barges.

Conditions of further transport

As is the case with transports of timber in terrain, the methods applied for further transport to the industries depend on the current conditions with respect to objects of transport, routes of transport, and means of transport. In addition there are biological and environmental matters to consider. The final choice of method is conditioned by economic circumstances.

Objects of further transport

Statements made under the heading Forwarding also apply in principle to further transport with respect to the objects of transport, their volume, weight and shape. Concerning volume, the value of concentrating large amounts of timber to the terminals should be emphasized, primarily with attention to efficient handling.

Weight is also important particularly at long transports. Weight of useless material e.g. water, bark etc should be avoided if possible.

Shape of timber is of great importance for the loading space at further transport. Reduced river drives and increased transport of timber on land has lead to an increased amount of standard length timber in some countries in order to better utilize the limited space on carriers.

Routes of further transport

Public roads, railways and rivers are primary routes of further transports of timber.

Forest roads

A separate system of forest roads between the harvest area and the public roads is needed for truck transports of timber. These forest roads make the harvest areas more accessible and they are therefore called access roads for the purpose of classification (Kolbas, 1983).

Expansion of the system of forest roads requires heavy expenditures that must be included in the total cost of transport as indirect travel costs. State grants may be obtained in some cases for the purpose of building forest roads (Coronan, 1983).

Forest roads are usually built not only to accommodate current load sizes and today's carriers but also to serve future road construction for bigger carriers with higher transport capacity. Thus, important factors are i.a. slopes, curves, width of road surface, maintenance, speed limits, location and design of terminals (Cf. Croisé, 1972).

Slopes

Steep slopes should be avoided when new forest roads are built. Slopes with a gradient exceeding 8 percent, requiring excessive hauling power, should not be allowed. This applies in particular to uphill slopes.

Curves

Depending on radius, curves influence both speed of travel and transport capacity. Transport is also affected by length of the curves and width of road in the curves. (Figure 192).

Width of road surface

Width of road surface should be adapted to the expected traffic flow. For light traffic it may be sufficient with turnouts at suitable intervals.

Meeting on the road must be possible for trucks with a width of 2.65 m.

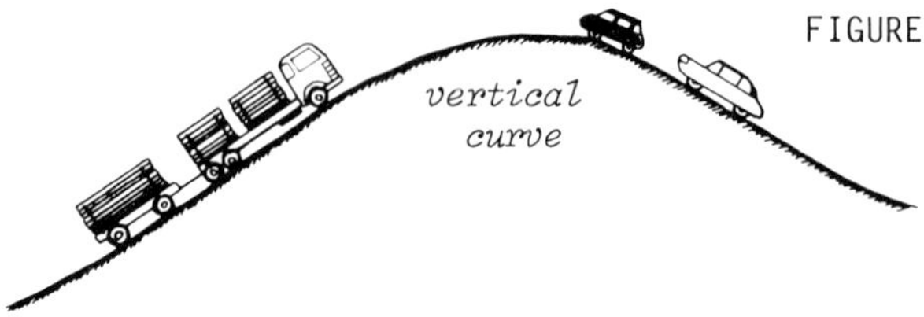

FIGURE 192. Vertical curves are particularly hazardous on account of poor visibility.

Maintenance of the roads

In periods of frost lift forest roads should be protected from heavy traffic. In winter, plowing and sanding in uphill slopes may be necessary. In summer the roads should be gravelled and graded as required.

Travel speed

Generally the following average speed is kept at truck transport of timber:

- main haul roads 12 - 15 km/h
- forest roads 32 km/h
- public roads 41 km/h

Location of terminals

Terminals for loading and unloading should have the same standards as the road system and they should be of sufficient size. Work at the terminals should be organized in order to reduce the waiting time for carriers as much a possible (Figure 193).

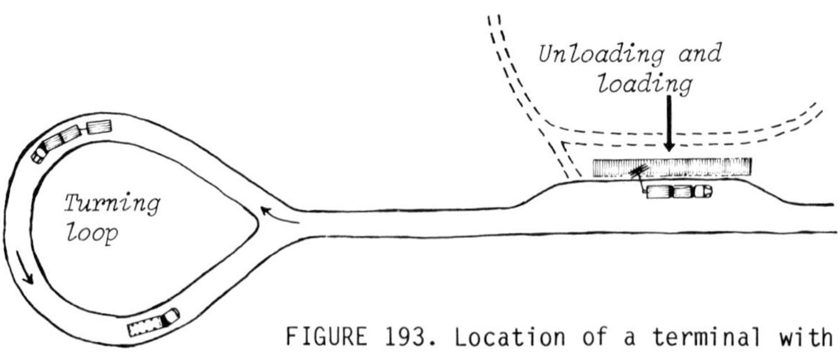

FIGURE 193. Location of a terminal with arrangements for turning

Means of transport

Trucks (lorries)

Trucks (lorries) are the dominant means of transport used for transports of timber on roads. Usually, the vehicle consists of a truck and a trailer. Depending on road standard, unit of handling or assortment of timber, there are various combinations of trucks and trailers (Figure 194).

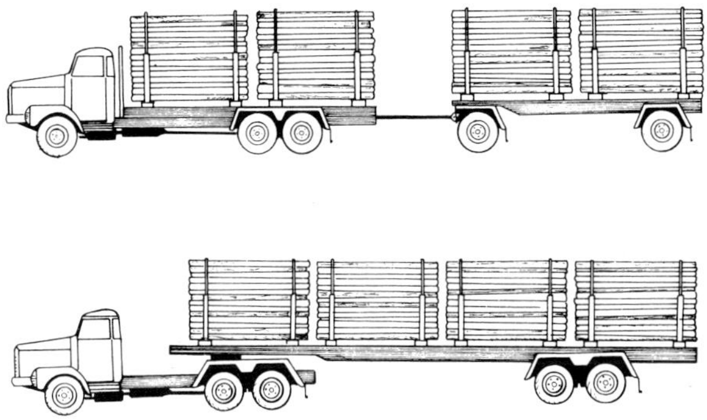

FIGURE 194. Bogie truck with 2-axle trailer (above) and truck with 2-axle semi-trailer.

Requirements concerning the truck

High load capacity is the first requirement of a modern timber truck. The basic hauling unit, therefore, is almost exclusively a 3-axle truck. Depending on transport distance and road class, it is combined with a 2-, 3-, or 4-axle trailer. Semi-trailers have also been introduced.

High hauling capability is the second requirement of the basic vehicle. To obtain sufficient hauling capability it is necessary to have an engine with high torque, a gear box with many shift positions and a powerful reverse gear (Figure 195).

The combination engine power-transmission is crucial for a good average speed on the road that will be travelled. The same principles apply as those discussed for transport in terrain. Thus, the objective is to attain a low cycle time per turn by means of high travel speed and short terminal time. Simultaneously, a high capacity should be ensured according to the following simple formula:

Output per time unit = number of turns per time unit x size of load per turn.

FIGURE 195. Performance of the truck depends on the hauling capability that can be achieved considering the pulling wheels of the vehicle and the conditon of the road surface.

Well organized loading and unloading is a third requirement of truck transport. Timber handling at loading and unloading is extremely strenuous work which now is mechanized (Figure 196).

FIGURE 196. Loading by means of a crane mounted on the truck.

Transport by trucks

Transport by trucks is virtually independent of seasonal changes in the weather conditions. It is flexible and also rapid. A transport of recently felled timber to industry can be direct. This reduces the costs of interest, timber defects and the handling in loading and unloading to a level that is rather proportional to the quantity of timber transported.

Transport can be done whether timber is bucked into various assortments or into random length of trunk sections, whole trunks or trees, directly to a processing or sorting depot at a central landing or to terminals near the industries.

FIGURE 197. The placement of a grapple loader:

A. Mounted in fixed position behind the driver's cabin.

B. Mounted on the rear part of the frame.

C. Mounted as a crane loader on a separate chassi.

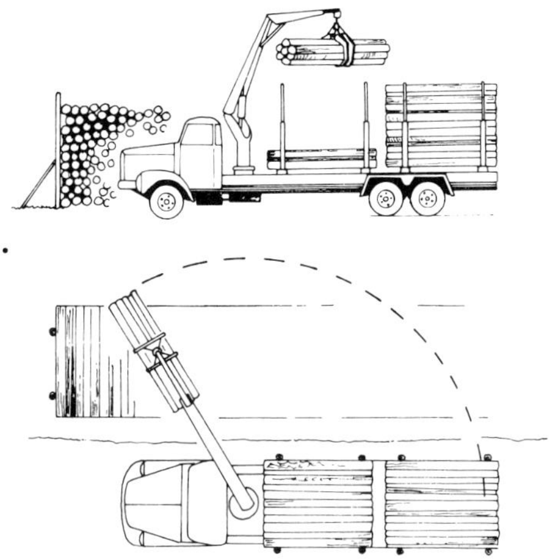

A.

B.

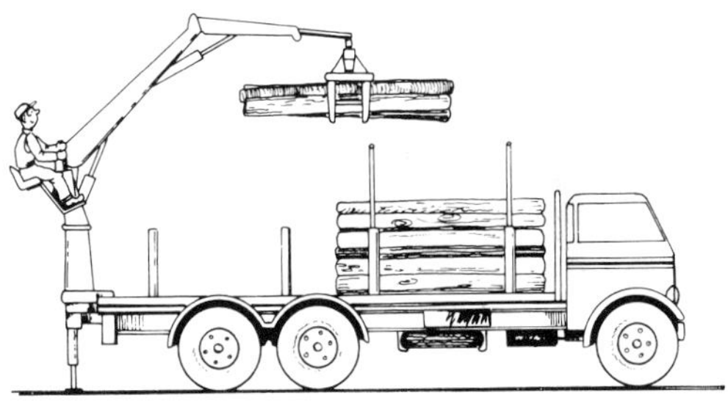

C.

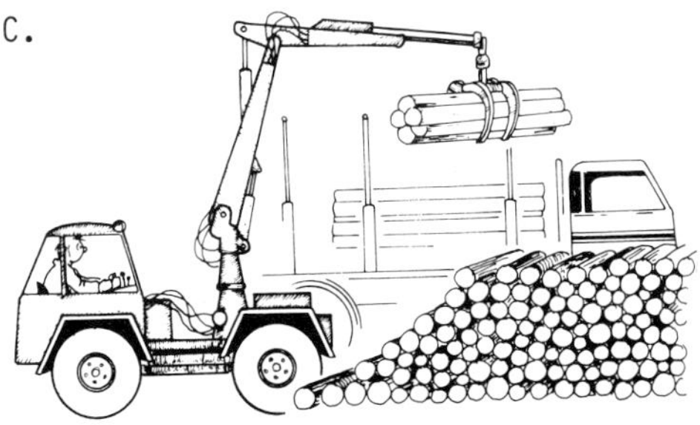

Loading

Loading can be done in several different ways depending on the technical means available and their placement in relation to the loading space of the truck (Figure 197).

Loading can also be done by means of fork loaders (rear or front) and cable cranes.

To facilitate the loading of a truck, it is very important that the unloading of timber from a tractor or some other vehicle used for transport in terrain is coordinated in a proper way.

The choice of loading method is influenced by the amount of timber. Loading for forwarding is different than loading for further transport.

Equipment with a very high capacity can be used efficiently when loading is done for further transports to industry. Unloading techniques may vary considerably between places of unloading (the industrial site, at railways or at riverside).

Separate loading machines are used primarily in intense shuttle traffic at landings with a large amount of timber or a heavy flow of timber. Truck mounted grapple loaders on hydraulic cranes are used at small landings with a limited flow of timber, e.g. a few truck loads per shift.

Unloading at an industrial site is usually done by means of gantry cranes or loading machines that are capable of lifting 10's of tonnes directly.

Unloading

New technical means of unloading have been developed (Figure 198).

Measurements and weighing of timber

Measurements at unloading are usually recorded in bulk volume of timber and supplemented by random sampling of the loads concerning density and quality.

Weighing of timber is occasionally the sole basis of payments for both harvesting and transport.

The combination of weighing with measurements occurs as well. This procedure is expected to increase in use if it can be accepted as a base for the economic transactions between agents and buyers.

Weighing of timber is considered to be a relatively fair method for payment of a transport in view of the great variations in wood density.

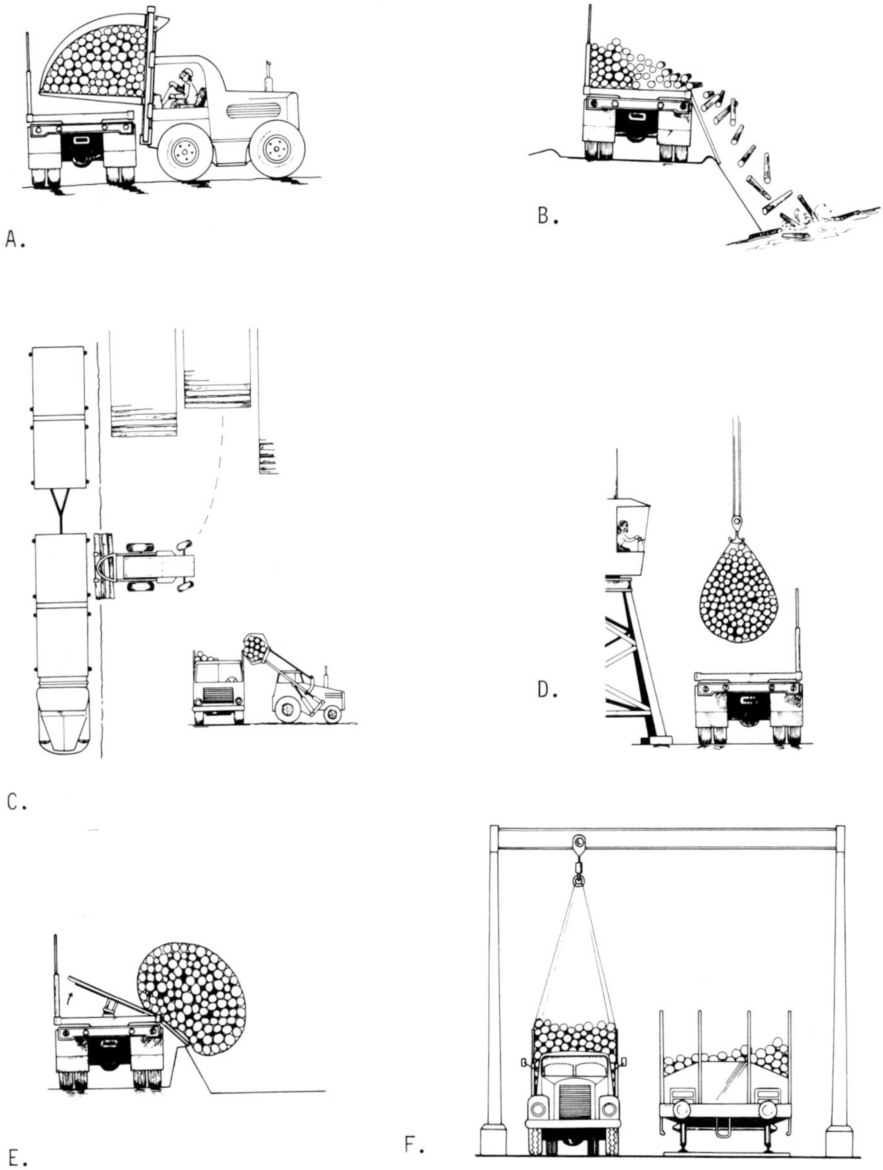

FIGURE 198. Various methods of unloading and reloading of timber at truck transports. A.: Rear fork loader, B.: Rush unloading (dumping), C.: Loading and unloading by means of front loader, D.: Loading or unloading by means of crane, E.: Unloading of bundles, F.: Unloading by means of gantry crane for loading onto railway carrier.

Combination truck and railway

It may be of interest in this context to review the system of further transport applied by a major forestry enterprise in which timber is first collected after truck transport to seven reloading stations within the former watershed. The truck transports replace the expensive and slow river drives in the tributaries. From the reloading stations the timber is transported by special trains to a major terminal located on the coast. Timber is then distributed by trucks from the terminal to various industries within the district.

This transport system has brought about a rationalization of the timber handling. The transition from timber transport by river drives in small tributaries to railways has also meant an improvement of the quality of certain end products e.g. newsprint, due to an improved supply of clean and sound timber.

Another advantage with this transport combination has been a considerably shorter time lapse between felling and processing in the mill with ensuing capital release. The end terminal covers an area of almost 20 hectares and the special trains are approximately 600 m long, comprising over 40 carriers, each containing two loading units, approximately 20 m^3 (500 cu.f) or 14 tonnes.

Since a train arrives every four hours throughout a 24-hour day, the rate of unloading is one carrier every five minutes, big fork loaders taking one loading unit at a time.

Specially designed trailers are used for the deliveries from the region of direct truck transports and from the terminal to the industries.

Timber is sorted at the terminal by means of an electrically maneuvered ramp into various assortments. Approximately 40 logs per minute pass by the operator who distributes the timber into 12 sections.

Railways

The conventional railway systems are increasingly being used for transports of timber in watersheds where river drives have been abandoned.

For further transport by railway timber trains are composed of specially equipped carriers. The carriers are adapted to timber of standard length or of restricted variation in length and designed for quick loading and unloading.

Rivers

In large regions rivers were once the original routes of timber transport. However, the importance of the rivers for transport of timber declined after World War II due to the rapid development of truck transport and the extensive construction of forest roads.

FIGURE 199. Loose timber in the river Dalälven, Sweden.

FIGURE 200. Movement of timber towards a dam and a chute in a small river. Getterån, North Sweden.

FIGURE 201. Rafting of timber on lake Päijänne in Finland.

Means of transport

The means of transport at river drives is the water which is regulated by structures of a permanent nature and by mobile equipment such as booms and boats with winch.

The permanent buildings designed for regulation of the level and flow of water (major rivers 200-300 m^3/s) consist of various types of dams as well as chutes made of timber, concrete or sheet metal, and steering screens, encasements etc.

The forward movement of timber along a waterway is achieved not only by means of water due to the gravitational pull but also by means of wind. Special auxiliary equipment is also used such as boats and rafts with winches, and ring booms for transport of timber, loose or bundled (Figures 199-201).

Usually, a waterway for floating of timber also includes large establishments for separation of the timber by various brands and owners, often adjacent to industry and, as a rule, at the mouth of the river.

River drive as a method of transport

River drive has always been an inexpensive and simple method of transport, having very large capacity for transports over long distances as well. However, the method has some disadvantages that are difficult to accept in today's situation.

River drives are relatively time consuming. From the moment when the trees are felled until the logs enter the sawmill or the chipper at the pulpmill, there may be a period of 2 years in extreme cases. In addition there are interest costs on the value of timber and the input of labour. An example from Sweden: Timber is felled in August 1981, transported to river in February 1982, transported by river drive in summer 1982, sorted and lifted into a timber yard in September 1982, kept in storage until August 1983 when it is brought into the mill and new timber is taken into the timber yard.

In addition, timber defects are caused by long time storage, and losses of approximately 2-3 percent, sometimes more, occur from sinkage during the river drive.

The value of water for generation of electricity is another factor of importance in closely regulated rivers. River drives require an ample supply of water and they are strongly seasonal.

The water supply in small tributaries is limited, often consisting of water from melting snow e.g. at drives in creeks and brooks. Extensive building of dams, chutes, guiding cribs etc. is then necessary in order to ensure a sufficient flow of water.

Small tributaries also require a close attention by a large labour force for the preservation of a continuous flow of water and timber. Because of the small quantities of timber, the high costs of labour have made the river drives in these locations very expensive and the small tributaries have been replaced with forest roads and trucks.

Relatively light construction and less labour input per km is required in the major rivers which have an ample supply of water. The large concentration of timber in a main river also contributes to a relatively low direct cost of transport.

Sorting and subsequent transport to the industries are additional stages of timber handling which have contributed to a switch to transports on land, primarily by trucks, for timber from the forests to the industries.

Methods of transport

Choice of method for further transport

The areas where trucks, railways or rivers are economically justified for transport of timber are determined by the total of direct and indirect costs of each method of transport.

In cases where the direct transport costs are decisive for the choice of method, transports on land by means of trucks or trucks + railways should primarily replace the more expensive river drives which are carried out in the tributaries of the main rivers.

If the relative influence of the indirect costs of transport per unit and a given length of road is great adjacent to a separation (sorting) establishment, transport on land should be justified in the neighborhood of the sorting establishment (industry).

Distribution of transports

As an example from the northern coniferous region may be mentioned that trucks delivered 70 percent of the timber to the industries while 25 percent was transported by rivers and 5 percent by the railways in 1950. Twenty years later the corresponding percentages were 87 percent, 3 percent and 10 percent, respectively.

Increase in transports by trucks depends on bigger loads, higher density of wood and longer distances of transport. During the same period engine power of the trucks had been raised from 87 kW to double that power. The road system also doubled.

Today more than 50 percent of the vehicle combinations have a gross weight of 36-41 tonnes at 10/16 tonnes of axle and bogie pressure. Approximately 20 percent of the vehicles have a permissible gross weight of 46 tonnes. With respect to length almost 40 percent of the vehicles measure between 20 m and 23 m.

Trends in further transports

Trends in further transports have been demonstrated in the preceding sections of this chapter. In summary it may be stated that the strong increase in truck transports has occured primarily at the expense of river drives and forwarding as a result of the expansion of the truck road systems in the forest operations.

Increased use of flatbeds for loading of timber on trucks appears likely.

Railway transports show a tendency to increase when waterways are abandoned (Hafner, Mihac, 1968).

Tree Harvesting Techniques applied in five basic methods

Various methods of harvesting

The next two chapters primarily deal with methods of harvesting and transport in various combinations which are presented as applicable. The matters concerning potential future methods of harvesting will be discussed with particular emphasis on thinning which is a prominent and important problem in the forests today. How should the stands be treated in order to prevent the development of slum areas with a poor future yield?

Thinning operations

Thinning operations are mostly motor-manual (semi-mechanized), the chain saws being the most important tools for felling, delimbing and bucking.

Intensified efforts are being made at developing more mechanized methods and systems for thinning operations. At early thinning there is a risk that processors and harvesters will damage the remaining trees. At late thinning, machines and methods designed for final harvest operations in old stands may often be suitable.

Final harvest operations

Tree harvesting in old stands has been subject to extensive rationalization and mechanization in the last three decades. The assortment ("shortwood") method dominates in many countries. Other methods such as the tree method, the tree part method and the tree length (trunk) method are quite common in other parts of the world. The latter methods have gained an increasing interest in recent years i.a. on account of the utilization of tree harvesting residues, tops and limbs, for pulping and fuel purposes. Chipping at the tree sites or at strip roads has been called the chip method, the methods of tree harvesting being named according to the form of timber being transported in terrain.

The following basic methods of harvesting have been distinguished:

1. Assortment methods
2. Tree length (trunk) methods
3. Tree methods
4. Tree part methods
5. Chip methods

Of these methods there are several varieties depending on degree of mechanization applied in the harvesting and transport operations.

1. Assortment ("short-wood") methods have a decentralized processing, scaling and bucking being carried out in the stands.
2. Tree length (trunk) methods are characterized by a decentralized delimbing and bucking.
3. Tree methods are characeterized by a centralized processing of the whole trees at landings on truck roads or at terminals for large areas.
4. Tree part methods are applied when the trees are bucked into feasible parts (sections) with limbs and, occasionally also the tops, are then transported to a central processing place or to industry. Parts suitable for sawing into lumber are distributed to sawmills and parts suitable for pulping to the pulp(board) mills while the limbs and tops are processed into fuel chips.
5. Chip methods consist of processing low quality or small timber and cleaning residues (small trees) by decentralized chipping in the stands or centralized chipping at terminals.

An important purpose of central processing at terminals is to achieve more correct and accurate scaling and, hence, improved utilization of the timber. It is also expected that all processing residues such as small trees, limbs and tops will be more efficiently utilized. However, terminals for centralized processing require ample space, a large supply of timber, and quick sorting techniques.

To utilize effectively the advantages of central processing, therefore, a well developed mobile equipment is needed. Terminals are located either at industries or separately. In countries where the assortment ("shortwood") method dominates the tree harvesting operations, the use of terminals for timber processing is less common than in countries where the tree method, the tree part method and the tree length (trunk) method are used more frequently.

The assortment (short wood) method

When the assortment (short wood) method is used the useful part of the felled tree is bucked at the stump into pieces of standard length or into logs of random length and forwarded over a short distance to a road.

The assortment method is applied to 95 percent of the total quantity of timber harvested in some northern countries. This proportion has previously been slightly lower but it now seems to be on the increase, on account of intensified mechanization in the young stands.

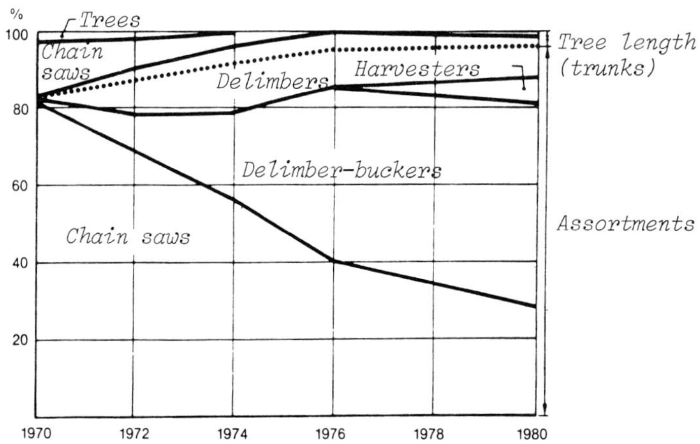

FIGURE 202. Example showing output of timber from various processes in the 1970's (Sweden) (Skogsarbeten, 1983).

The assortment method is primarily used in regions or countries with industrial forestry where the stands consist of a limited number of species and where the trees are relatively small and suitable for pulping purposes. The terrain, which is rather easy and plane, can be travelled by vehicles almost everywhere and the distances to the wood processing plants or to the industrial centres are rather long. The tree method and the tree length (trunk) method are usually applied in countries where terrain is alpine, mountaineous and steep. Then they are largely combined with various systems of cableways (Samset, 1979).

Various systems used in the alpine countries are also well described in Mechanozovani Transport Drveta (Hafner and Mihac, 1968). Elsewhere, skidders, grapple skidders and winch cranes are used for transport of timber downhill to a graded landing for further processing or the timber may be loaded onto trucks for further transport to some major, centrally located place of processing.

The tree method and the tree length (trunk) method are used commonly in regions with a large variety of species or particularly valuable trees that require careful scaling. The tree method and the tree part method are economically advantageous in areas where the tree harvesting residues, limbs and tops, are valuable for fuel purposes. In forest regions which are located in the vicinity of sawmills or pulpmills, harvesting by the tree method, the tree part method or the tree length (trunk) method may be the most suitable methods. the choice of tree harvesting method for each location must be carefully taking into account all important advantages and disadvantages with the various methods.

Semi-mechanized assortment method

The assortment method applied most commonly today is semi-mechanized, consisting of motor-manual felling-processing and mechanized transport in terrain. The method is outlined in Figure 203.

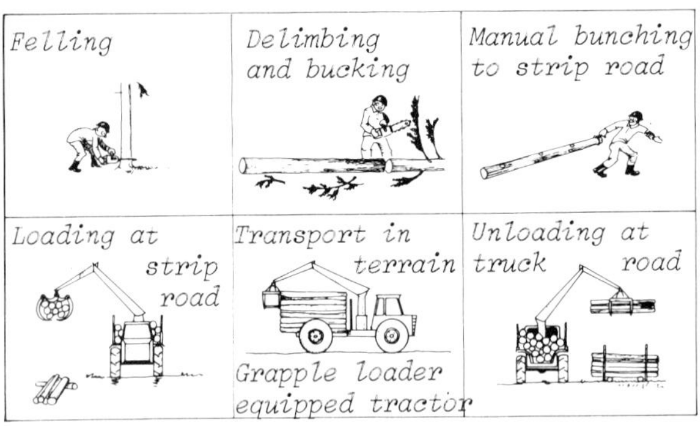

FIGURE 203. Assortment method.

Entirely mechanized assortment method

Originally the chain saw was used for all felling in the assortment method. However, in the beginning of 1970 special felling machines were introduced. Entirely mechanized delimbing and bucking operations are now being used more and more, either by means of delimber-bunchers and manual bucking or by means of processing machines for both delimbing and bucking at strip roads. More recently a bucking method by means of a chainsaw built into the grapple of the crane has been added.

As mentioned previously, a relatively large number of modern processors such as delimber-bunchers and delimber-bucking-bunchers have been introduced. The use of these machines continued at an increasing rate, gradually leading to highly mechanized assortment methods. Due to a high productivity and because of high capital costs each machine must be fully utilized in order to give a high annual production. The output of the processing machines, therefore, soon manifested itself in the statistics (Figure 202).

The tree length (trunk) method

The tree length (trunk) method is applied when entire, delimbed trunks, often topped, are transported over short or long distances. The method also includes several varieties i.a. a semi-mechanized combination where felling, delimbing and bucking is done by means of chain saws and transport by means of skidders (Figure 204).

FIGURE 204. The semi-mechanized tree length method using a skidder equipped with winch.

Entirely mechanized tree length method

A mechanized method with motor-manual felling and mechanized delimbing, bucking and transport was expected to produce a major part of the total quantity of timber harvested by the tree length method in the 70's.

Today mechanized delimbing is done primarily by means of a machine, the delimbing output of which may amount to approximately 90 percent of the total quantity of timber produced by the tree length method.

Transport in terrain is carried out either by means of special tractors equipped with winches, so-called winch skidders, or by means of tractors equipped with clam banks and grapple loaders, so-called clam bank skidders, which are used for transport of the trunks bunched after passing through the processing machine. A grapple skidder included in this method would have a too low transport capacity compared with that of a skidder with clam bank.

When trunks are skidded by means of winch, the top ends are usually hauled first primarily because it is easy at directed felling to choke and bunch the load with top ends pointed towards the winch. Skidders with winch have a sufficiently high hauling capability for this transport which requires a high capability because of the high resistance to skidding.

When skidding is done by means of clam banks, which is increasingly being used, or by means of grapple skidders, the butt end is hauled first.

Hauling with butt end first is often preferred when small skidders with low hauling capability are used and in difficult terrain, e.g. steep slopes and soft ground. On extensive plains and on soft ground both trees and trunks are often transported with butt end first.

Bucking of the trunks can be done either on the upper or the lower landing or at a place of processing.

Bucking by means of chain saws is currently estimated to be done for a large part of the timber processed. The mechanized bucking on the upper and the lower landings is done by means of bucking saws of various types. The bucking saws for upper landings are mobile units.

Lower landings and industrial sites with a large flow of timber are equipped with semi-stationary and permanent establishments, respectively, for bucking and sorting which is more or less automatic.

Transport of trunks between the upper landing and the lower landing, or industrial site, is done by means of specially equipped trucks that have i.a. separate grapple cranes for loading of trunks (Figure 205).

The tree method

The tree method, previously often called the whole tree method, is applied when felled not delimbed trees, bucked at the stump and sometimes also at the top, are transported over short or long distances.

FIGURE 205. Loading of trunks onto truck.

The *tree method* and the *tree length (trunk) method* have long been applied in several countries. A flexible and efficient utilization of the timber resource will require improved techniques in these methods and in the more recently developed tree part method.

Transport of trees, tree parts and fuelwood (harvesting residues) requires specially adapted equipment and new techniques for loading and unloading. This applies in principle to transport by trucks as well as transport by railway. Knowledge of and experience with transport of trees and tree length timber is deficient in countries where the assortment ("shortwood") method dominates tree harvesting work. This lack of knowledge must be filled before the tree method and the tree length (trunk) method can be expected to gain a more general application. Improved utilization of the timber resource also requires that the size and quality of the timber can be measured more accurately. In spite of requirements for high production the machines for delimbing and bucking must function with great accuracy. Machines for sorting or separation of bundles into individual logs must operate with great precision and a low frequency of disruptions. Demands for high quality timber will require improved scaling and measurement procedures.

Scaling of conventional assortments is a long used technique. However, measurements of stumps, wood fuel, whole trees and tree parts with limbs are relatively new activities with problems that are to be solved for tree harvesting systems with terminals.

The tree method is applied to a large extent in USSR, where timber depots are being used. At the depots all parts of the trees are utilized, even limbs e.g. for the manufacture of boards.

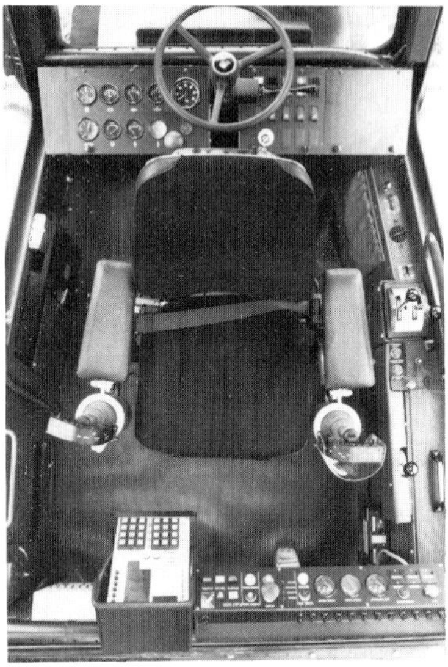

FIGURE 206. The operator's cabin with maneuvering panel. Automatic scaling requires well developed recording mechanisms.

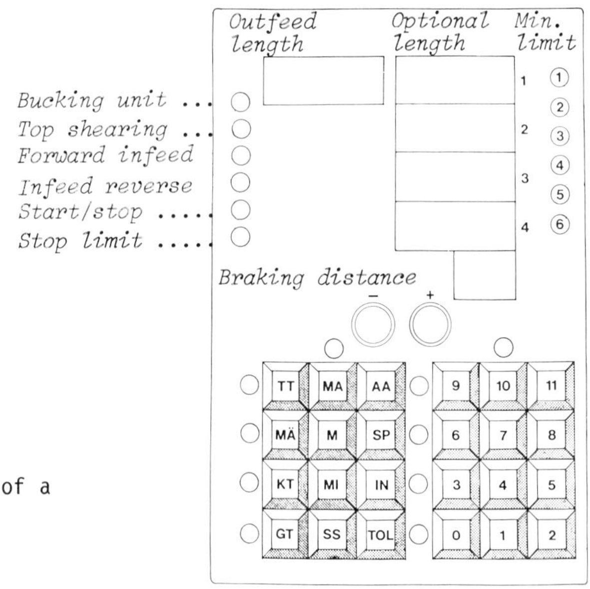

FIGURE 207. Maneuvering panel of a micro-processor.

In Sweden only 1 percent of all timber is harvested by the tree method. The major part of this amount is done by means of delimbing depots. The method appears to be declining in use for reasons mentioned previously while it seems to gain ground when trees are used completely for energy purposes.

The partial operations of the tree method, delimbing and bucking, are usually entirely mechanized.

Skidding to the upper landing is done after mechanized felling by means of feller-skidders equipped with clam bunks.

Skidding after manual felling is often done by means of skidders equipped with winch and chokers. Butt ends are usually hauled first.

The tree method may also include the use of bunch delimbers on landings. The trees are transported by means of feller-skidders which can deliver suitable bunches directly at the delimber. This is feasible at harvesting of small timber.

The tree length method and the tree method are generally sensitive to stoppage in the chain of production. The methods are used under the condition that timber is to be delivered continuously to the industries from the processing places. This condition may lead to problems i.a. in primary timber production.

The tree part method

In recent years the tree part method has been tested in order to coordinate the utilization of industrial wood and fuelwood.

Examples of tree part methods

A. Thinning operations

Step:

1. Felling by means of chain saws.
2. Bunching by means of winch or a crane with long boom. An open net of strip roads is desirable.
3. Before loading, the unsorted trees are bucked into 5.0 m - 5.5 m length by means of a grapple saw.
4. Forwarding preferably by a small 8-wheel unit requiring narrow strip roads only and causing slight damages to the ground.
5. Timber is put into relatively high piles at the truck road.
6. Timber from the tree part method is transported by truck with trailer (high sides). Loading capacity available for a normal vehicle may be 100 m^3. Special compactors may be used for improving the solid wood content of the loads at long transport.
7. Net weight of load at industry is registered and its moisture content

is determined by random sampling.

8. The timber is unloaded into an intermediate storage space and sorted roughly into tree classes of species mixture.
9. Bundles of trees are transferred from the intermediate storage space by means of trucks to the infeed table. The trees are lifted into a chamber for heating and thawing of snow and ice in winter.
10. The trees are dropped from the chamber into a specially designed, horizontal and conic thumbler which is 30 m long and 3.8 m - 5 m in diameter (Figure 209).
11. Foliage is removed before delimbing at the midsection of the thumbler. After debarking at the small end of the thumbler, the timber is washed and chipped.
12. Chips are screened and stored in piles and ready for the digesters in the kraft pulpmill.
13. The thumbler separates approximately 40 percent of the biomass which is stored together with bark for fuel purposes.
14. The wood fuel is transported from storage to various consumers.

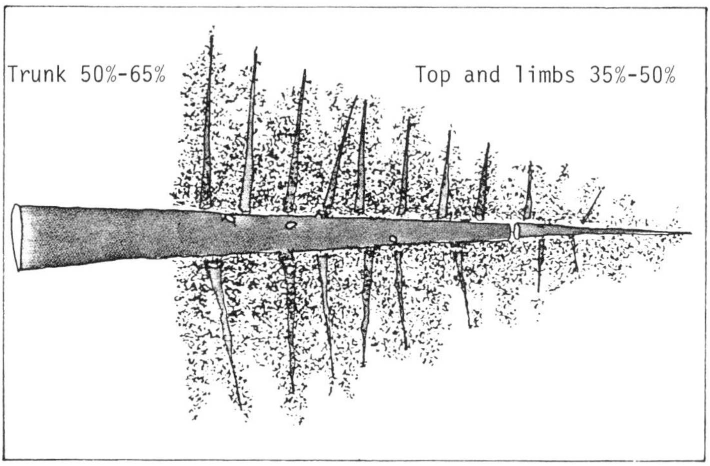

FIGURE 208. Biomass distribution in a tree from thinning.

In recent years the tree part method has been tested in order to coordinate the utilization of industrial wood and fuelwood.

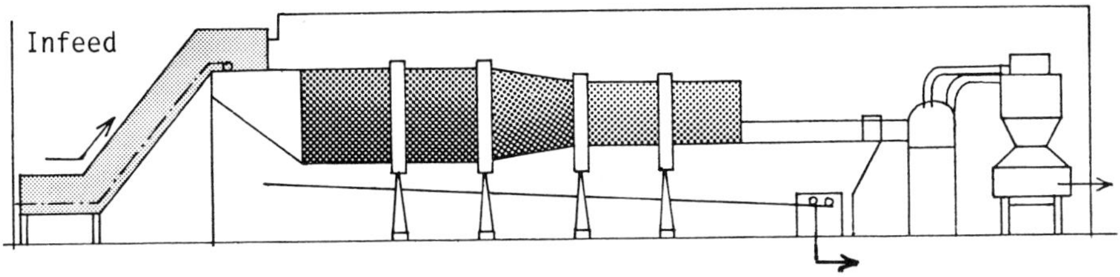

FIGURE 209. Principle outline of a tree rinsing unit (Billerud, 1983).

B. Final harvest operations

Step: (two alternative examples)

1a. Felling of trees in swaths by means of chain saws and immediate processing of the timber by means of chain saws.

1b. Mechanized felling and immediate processing of the timber by means of chain saws.

2. Bucking by means of a grapple saw mounted on a forwarder.
3. Immediate loading of the forwarder, tree parts separated from bucked timber.
4. Transport to truck road for further transport of tree parts to a central place for processing where the tree parts are delimbed in bunches (several tree parts at a time).
5. Timber is processed for pulping and the limbs are used as fuel at the pulpmill.
6. Sawlogs are transported immediately to a sawmill.

(Larsson et al, 1983).

The best location for a stationary or semi-mobile delimber is adjacent to pulp industry or a sawmill. However, it is not clear whether the tree part method in the forest is more advantageous than the ordinary assortment method (Gustafsson and Laestadius, 1983).

The chip method

Chipping of trees from cleaning at truck road

Chipping outside the stands at strip roads or truck roads is usually done by means of a tractor equipped with winch or grapple for skidding, and/or cart. Vehicles equipped with winch can operate over a distance up to 60 m between the strip roads where the trees are chipped (Figure 210).

If a forwarder is used for transport of trees from the cleaned stand to truck road or landing, the strip roads may be laid out only 20 mn apart. The trees bunched into piles are loaded by means of the grapple.

Comparing the three methods of chipping, viz. in the stand, at strip roads after collection of the trees by means of a small tractor (open net of strip roads), or after transport by forwarder (dense net of strip roads), it appears that chipping in the stand is more profitable. Output is then high in relation to the level of machine costs. Damages to the site and the remaining trees can be kept slight. However, differences in output between the three methods seem to be rather small.

FIGURE 210. Chipping of small trees in stand.

Transport of chips to the consumer

Chips can be transported in high tipping containers or in large sacks. The system with large sacks provides a new alternative for handling and distribution of fuel chips or green chips for further processing into pellets. The volume per sack is 2.5 m^3, its length about 3 m.

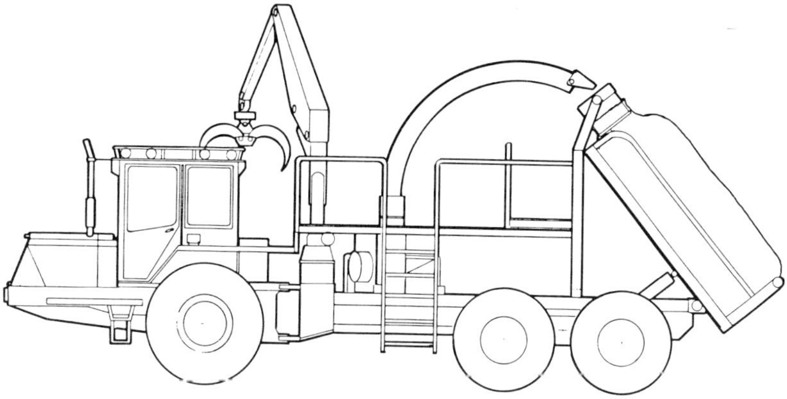

FIGURE 211. Chipping into large sacks.

The system with large sacks provides a choice of:

- various systems of handling in further transport and distribution
- transport on forwarder or on standard trucks.

- no obstructing separations inside the container
- flexibility at combined transport
- automatic locking of side panels
- double acting cylinders for tipping.

FIGURE 212. Example of trailer for transport of chips - a 4-axle trailer.

Transport of residues for chipping at industry

FIGURE 213. Trailer combination - 4-axle - for transport of limbs and limby timber to industry.

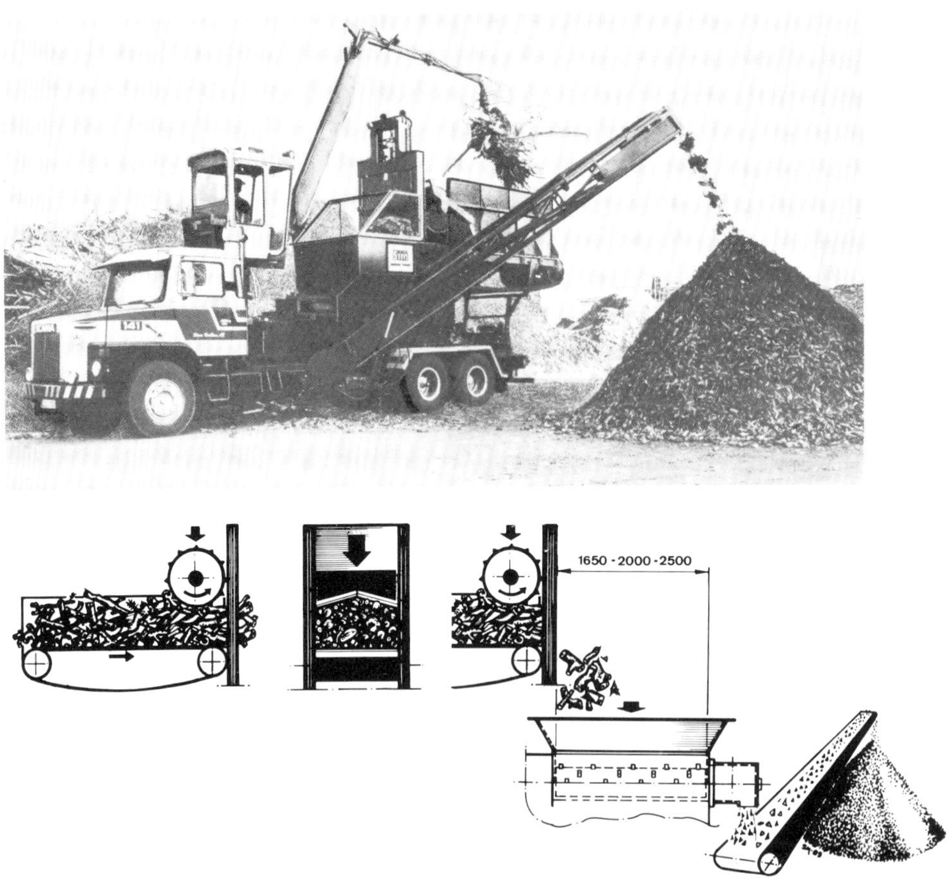

FIGURE 214. Example of wood crusher for conversion of forest residues and industrial wood waste into valuable fuel.

There are three different principal types of wood crushers:

- stationary types
- semi-stationary types
- mobile types

in various sizes with single or double crushing rollers.

Chipping of stumps

For harvesting of stumps which may become of greater interest at shortage of timber, more rational techniques must be developed in order to improve the potential of this type of timber harvest. At present the stumpwood extracted by means of units mounted on excavators is often mixed with rocks, gravel and soil. The separation of impurities by crushing and washing before the stumpwood can be chipped for pulping or fuel purposes is fraught with difficulties and expensive. Extraction of stumps must also be judged from environmental and plant nutrition points of view.

FIGURE 215. Stump extractor in operation.

Chipping integrated with the tree method and the tree part method

The _tree method_ is applied when the whole trees above the ground are transported from the stands in one piece.

The tree part method is applied when the trunk, bucked into sections with limbs remaining, is transported from the stands.

Processing of trees or tree parts is done in a subsequent operation at truck road or at industry. These methods provide a concentrated production of fuel from limbs, tops and small trees i.e. an additional supply of chips corresponding to 20 percent - 40 percent of the tree volume above the ground.

The tree method and the tree part method applied at thinning operations

A. Equipment for the tree method (Example)

Base machine (a small tractor), winch and hydraulic grapple for skidding.

FIGURE 216. Tree harvesting by means of grapple for skidding.

The method is demonstrated in the figures.

1. Felling in swaths at 30°-45° angle to the strip roads, butt ends toward

the roads.

2. Coupling at the butt end, winching directly into the open grapple. Each winch load contains 3 - 7 trees attached along the winch cable.
3. Skidding to landing. Load contains 15 - 20 trees.

FIGURE 217. A Danish attachment to the tractor, lifting the tree by means of a telescopic boom, turning the tree in vertical position and laying it on the clam bunk.

B. Equipment for the tree part method (Example)

Base machine, winch for skidding, grapple loader, and bogie cart.

The trees are pulled by the winch to the strip road and bucked into sections which are loaded by means of the grapple loader onto a bogie cart, butt ends on the cart. The sections (tree parts) are then transported by skidding to a landing. This method requires a bogie cart with great stability (Liss & Risberg, 1983).

1. Motor-manual felling

2. Mechanized felling and bunching

3. Mechanized felling, bunching and skidding

4. Mechanized felling, bucking and delimbing (harvester)

5. Mechanized felling, bucking, delimbing and forwarding

6. Motor-manual delimbing

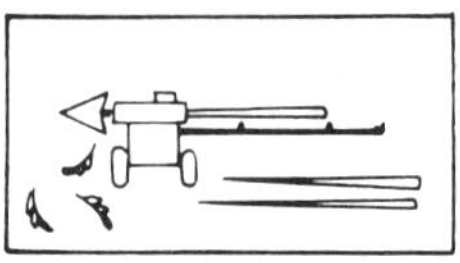

7. Mechanized delimbing

8. Motor-manual bucking of tree length

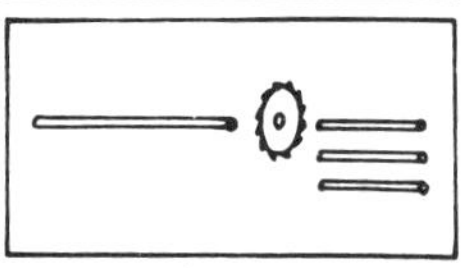

9. Mechanized bucking of tree length

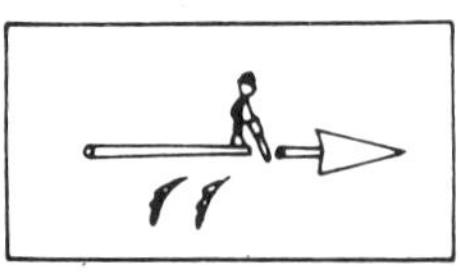

10. Motor-manual delimbing and bucking of tree

FIGURE 218. Denotations for various partial operations.

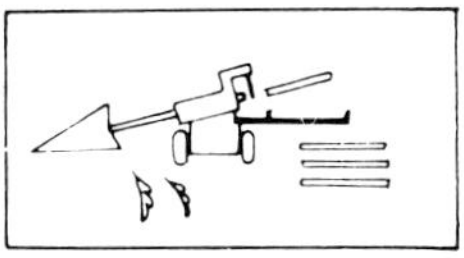

11. Mechanized delimbing and bucking of trees

12. Mechanized delimbing and bucking of bundles

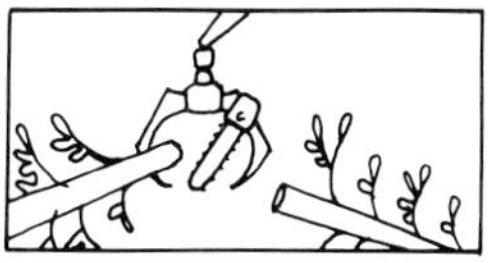

13. Mechanized bucking of trees into tree-parts

14. Chipping of small trees at strip road

15. Forwarding of assortments (shortwood)

16. Skidding of tree length trunks

17. Skidding of trees

18. Transport of tree-parts or small trees along strip road

19. Transport of chips along strip road

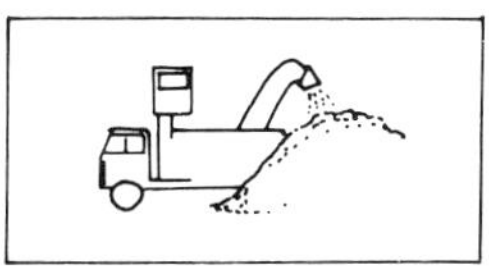

20. Chipping at upper landing

21. Further transport of assortments (short-wood) by trucks

22. Further transport of tree length trunks by trucks

23. Further transport of trees by trucks

24. Further transport of tree-parts or small trees by trucks

25. Loading of chips

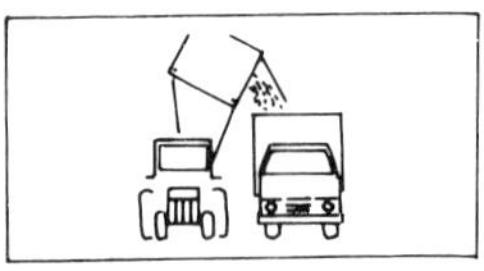

26. Reloading of chips

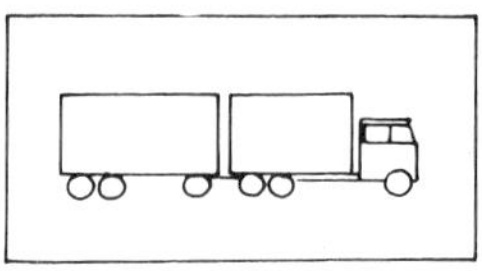

27. Transport of chips by trucks to lower landing or industry

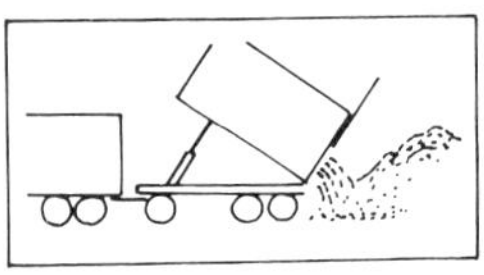

28. Unloading of chips at lower landing

29. Chipping at lower landing

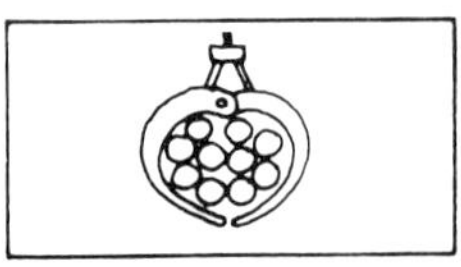

30. Loading

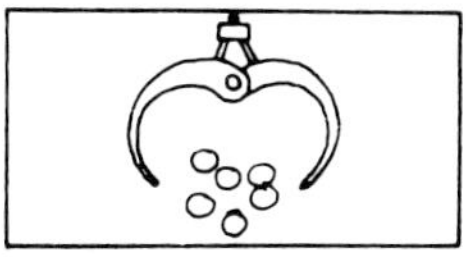

31. Unloading

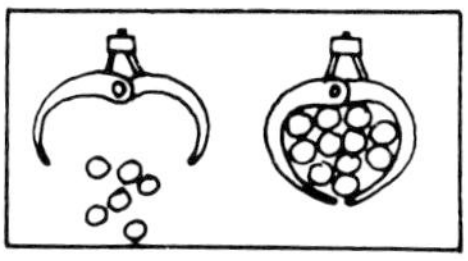

32. Reloading

Various degrees of mechanization

The methods of harvesting represent various degrees of mechanization from muscular to entirely mechanized forms of work. A large number of harvesting methods are theoretically possible in today's situation. The methods may be presented by means of a system of denotations.

Numbers can be used to denote the harvesting operations which are carried out in various places in the chain of transport e.g. number 1 for felling, number 6 for delimbing, number 8 for bucking etc. These numbers are then placed in various boxes representing different places of processing.

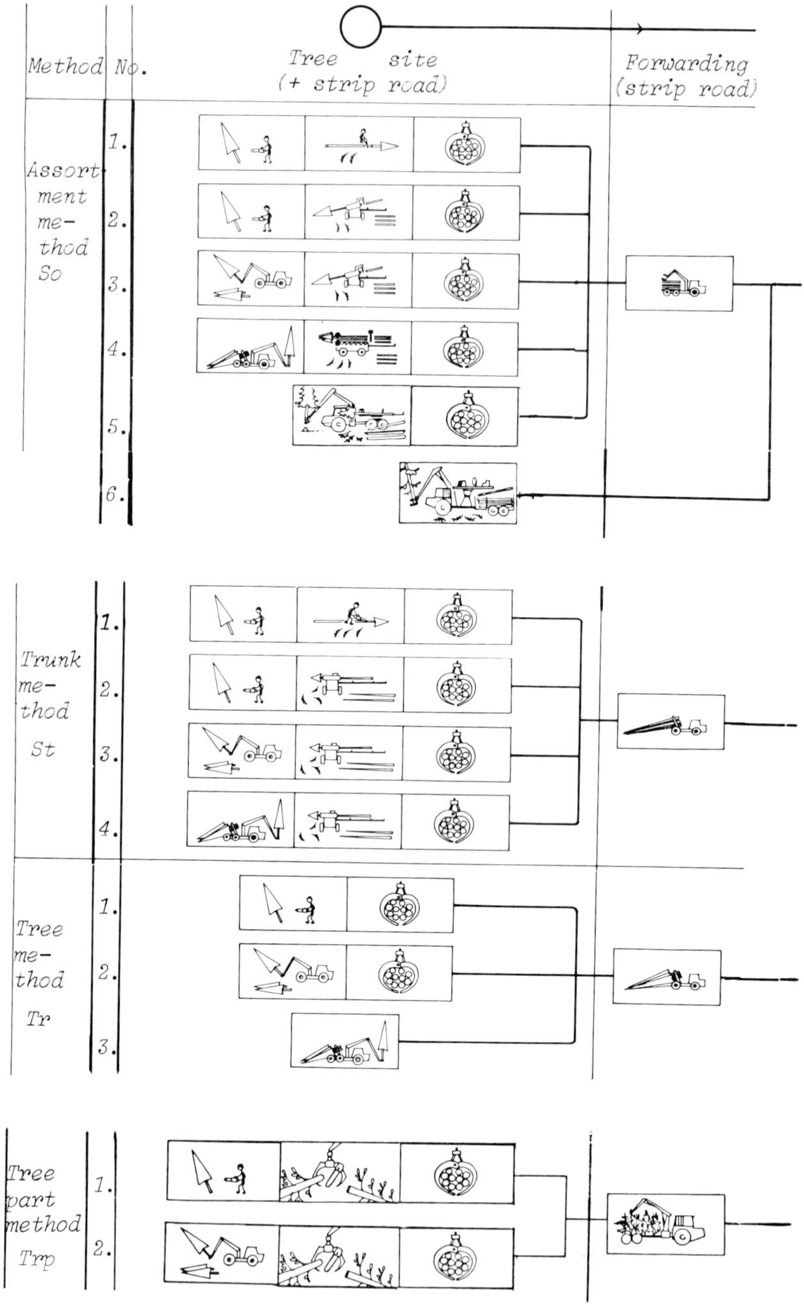

FIGURE 219. The most common methods of harvesting.

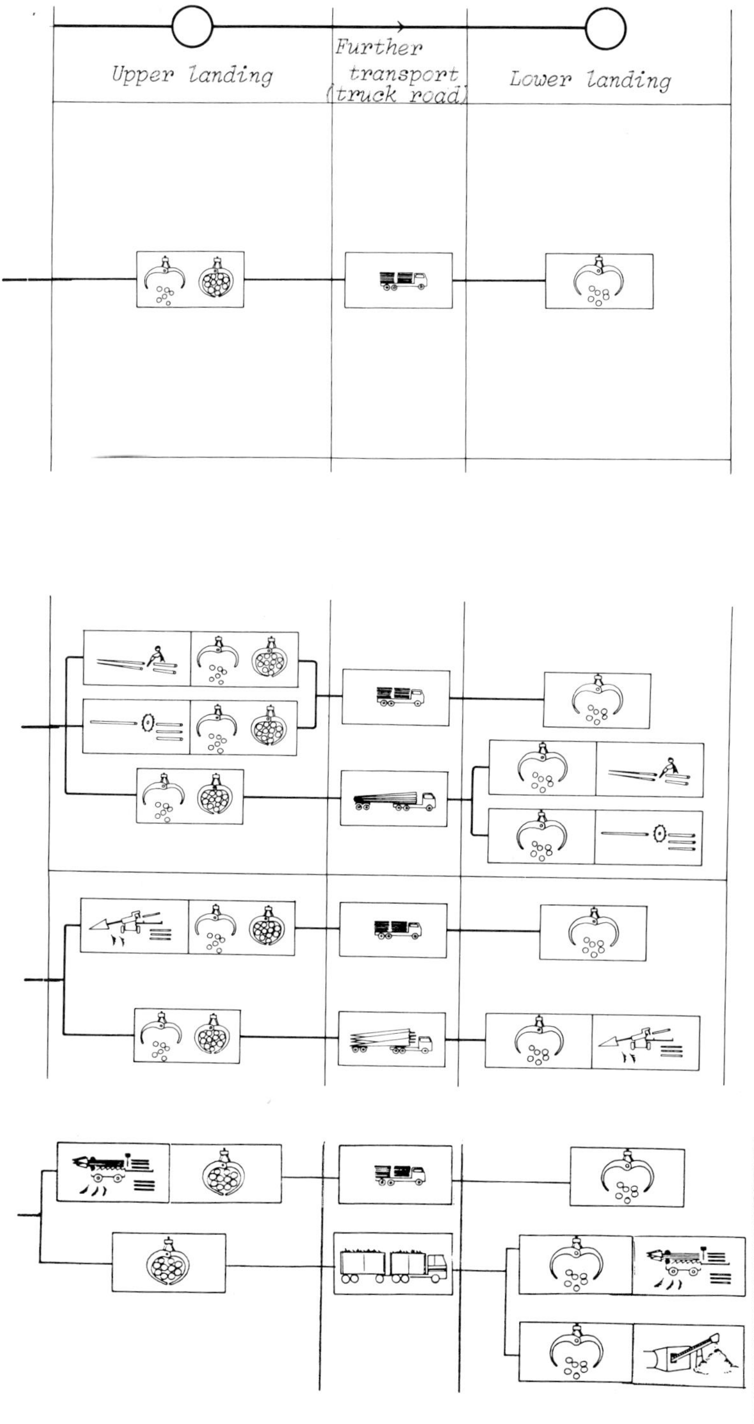
Upper landing
Further transport (truck road)
Lower landing

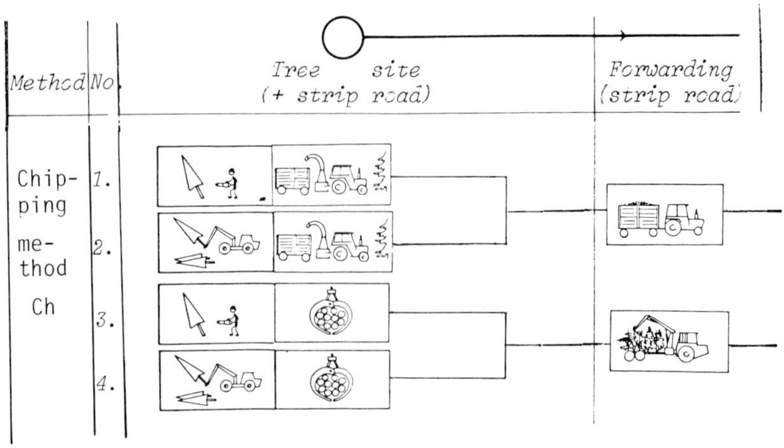

Partial operations

This outline can also be extended by denotations for the degree of mechanization in the partial operations and for the functions of the technical means in the various methods. Thus, more than 60 different varieties of harvesting systems can be composed. A comparison between various current systems can be made by evaluating various machines and methods with respect to performance and costs i.a. by means of established time formulae in a simulation approach (Newnham, 1972).

The most important partial operations applied in today's harvesting and transport systems are presented by means of denotations compiled in Figure 218.

The partial operations are combined in 19 different methods of harvesting in Figure 219. Six assortment methods, four trunk methods, three tree methods, two tree parts methods and four chipping methods with different degrees of mechanization are presented. All transports are entirely mechanized and the machine input in harvesting, therefore, is decisive for the degree of mechanization in the five principal methods.

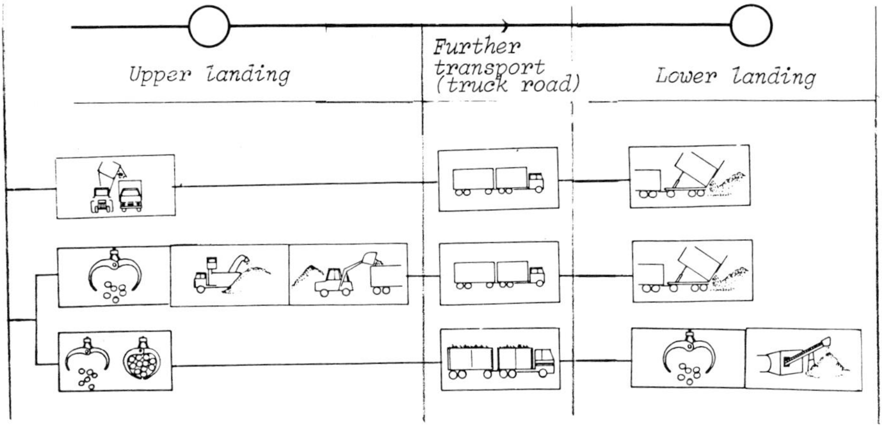

Choice of harvesting method

Factors influencing the choice of harvesting method

The following factors are of importance for the choice of harvesting method:

1. Form of harvest, thinning or final harvest
2. Volume of timber per hectare and total volume
3. Tree sizes
4. Limbs
5. Terrain conditions
6. Transport distances and road standards
7. Terminals
8. Amount of snow or rain

Methods of harvesting in thinning

Conventional thinning removing a low volume, approximately 50 m^3 per hectare or less, and small trees is dominated by the assortment method, So

1 in Figure 219. This method consists in harvesting by motor-manual, directed felling and manual bunching of timber into piles along the strip roads.

A variety of this method consists in bunching of timber in the stands by means of tractor mounted or portable winches with cables. This variety allows a considerably more open system of strip roads and the frequency of damages to the remaining trees is reduced (Putkisto, 1970).

The feller-driver operation is another form of harvesting used by forest owners in small operations by means of simple farm tractors equipped with winch. All work is carried out by one man, usually the forest owner himself.

The problems encountered in entirely mechanized thinning have been discussed in some detail in the preceding chapters. Various potential systems will be presented in this chapter (Cf. Cornides, 1973 and Eisenhauer, 1981).

Thinning under very difficult conditions, e.g. in rocky terrain, can be done by means of winches and a high cableway, by the tree length method, or the tree method. Subsequent processing can be carried out either at the upper landing or at the lower landing. See form of methods ST 1 or Tr 1, for trunks and trees, respectively, in Figure 219.

Thinning operations

Thinning operations are largely motor-manual. Chain saws are used for felling, delimbing and bucking. The trees are bunched manually to the strip roads for transport by forwarder to a truck road. Bunching may also be done by means of winches or cranes with long booms. (Winching can actually be considered motor-manual).

Development of systems for mechanized thinning operations has proceeded toward simplified and smaller machines. Often the felling is first done by means of chain saws and winching has been added for increased concentration of timber at the strip roads. The degree of mechanization is rather low and the cost of operation is high. To obtain a higher concentration of timber, cranes with long booms have been introduced. Mechanized felling easily causes damages to the trunks of remaining trees. New harvesters for thinning operations using new components for felling and bunching are subject to a rapid development.

When the tree part method is applied in thinning operations, an adaptation of forwarders and grapple saws or a development of entirely new types of machines will be necessary for e.g. the combinations of felling and skidding or felling, bucking and forwarding (Arvidsson et al, 1983).

For further transport of trees or tree parts from thinning operations, techniques currently used will require improvements.

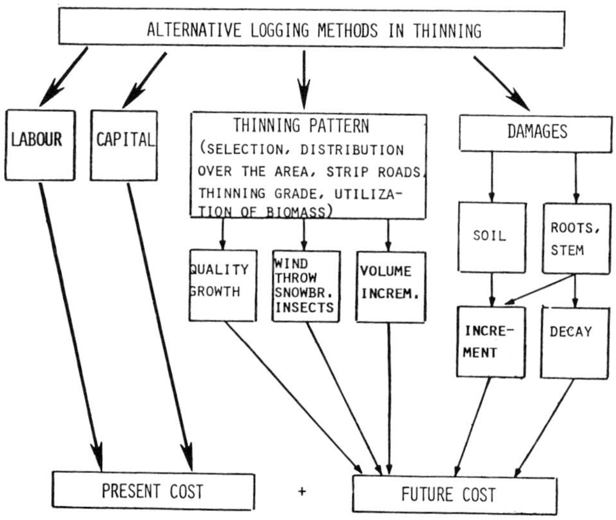

FIGURE 220. A model for the valuation of thinning systems (thicker arrows refer to "activities" within a project "Thinning Techniques" (Arvidsson et al, 1983).

Planned motor-manual felling in thinning operations

1. Conventional thinning operations

Distance between the strip roads should be at least 20 m. The strip roads should be laid out as straight as possible with a width of 4 m, in curves and slopes 5 m. A zone of 3 m width should be reserved for storage of timber.

2. Thinning operations with winch

Distance between the strip roads should be approximately 100 m. The strip roads should be located on the most suitable ground. Width of the strip roads should be 5 m and the storage zones on each side of the strip roads should be 6 m wide.

Principles of felling in conventional thinning operations

- The trees are felled in order to provide a suitable working height.
- The trees are felled into areas already cut so that timber is in or close to the zones of storage while the limbs are put in the strip roads or in the zone between the strip roads.
- Trees standing in the road location are felled along the road toward the thinned part of the stand so that the timber will be placed close to the storage zone and the limbs will stay in the road.
- Trees standing in the storage zones are felled at an oblique angle to the strip road or into the zone between the roads toward the areas which have already been thinned. Timber will be placed in the storage zone and limbs will be put in the road or in the zone between the roads.
- Trees standing outside the storage zones are felled at an oblique angle toward the strip roads and directed toward the areas that have already been thinned. Timber will then be easily placed in or adjacent to the storage zones while the limbs can be put in the strip roads or outside the storage zones.

The limbs should be put in the road for the purpose of providing an improved carrying capacity and protection of the ground at the subsequent transport of timber by means of forwarders.

Principles of felling in thinning operations with winch

- Trees in the strip roads are felled so that as much limbs as possible can be put in the road.
- Trees standing in the storage zones, which are thinned conventionally, are felled so that the limbs will be left in the road while the timber can be placed in the storage zones.
- Trees standing outside the storage zones are felled toward the strip roads, tops concentrated in swaths for winching.
- The felled trees are delimbed and scaled with marked points of bucking on two sides of the trunks.
- The trees are pulled by means of winch toward the strip road, the nearby trunks taken first.

- The trees are aligned along the strip road before bucking
- The swaths of winching are laid out at an angle of approximately 45° to the strip roads (fishbone pattern).

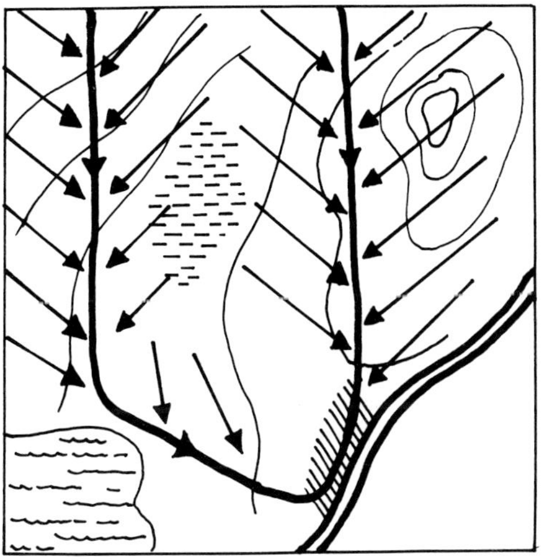

FIGURE 221. Swaths of winching in fishbone pattern.

3. Thinning operation with clam harvester mounted on a crane

The beginning of the 1980's saw a rapid development of felling units which, in combinations with various types of infeed mechanisms, spike rollers or rubber covered rollers and delimbing knives, could also be used for bucking and other processing. This combination has been called clam harvesters. They are designed for mounting on the crane of a base machine e.g. a forwarder. This felling unit can be used in cleaning, thinning and final harvest operations. The main part of the unit is a built-in swinging saw blade with a hydraulically powered chain. Hydraulic clipping-shearing tools may also be used, particularly for small trees. For felling and bunching of industrial timber by means of an accumulator, the unit may also be equipped with a heavy saw that is able to cut trees with a diameter of up to 40 cm, as well as undergrowth and bushes.

The 1983 models of clam harvesters vary in weight between 225 kg and 850 kg. The maximum diameter of the tree at stump height that can be cut by this machine varies between 20 cm and 50 cm depending on the mode of application. In addition to pairs of spike rollers or rubber covered rollers, stepwise feed mechanisms may also be used in the units.

Equipment for measurements of length are mounted on most of the clam harvesters. This will allow automatic or optional bucking. The clam harvesters mounted on a base machine require an engine output of at least 52 kW - 105 kW. It is also necessary that the lifting capability of the crane is at least 3.6 tonne-m - 9.0 tonne-m depending on the weight of the unit.

The modern clam harvesters provide opportunities for several new methods of felling and further processing. Development objective of the clam harvester was to utilize better the existing machines and to lower the high capital costs of the tree harvesting operations. Primarily they are designed to be used for felling of small trees e.g. in thinning operations. When the harvesters are used for felling of large trees in final harvest operations, there are problems with stability and the cranes are subject to great stress because of the heavier units necessary. However, there are clam harvesters developed for final harvest operations and late thinning operations and they can be used as equipment for delimbing and bucking. Weight 650 - 850 kg.

When clam harvesters are used, the capital costs can be kept relatively low, which is important in thinning operations. Since the mode of production is a coupling of the partial operations in a series, output of timber is low per unit of time. Contrasting this mode of production is the use of a harvester with fixed delimber-bucker mounted on the base machine and a felling head with accumulator. The partial operations carried out by the harvester can be coupled parallel to each other e.g. 3-4 trees can be collected while other trees are processed automatically e.g. delimbed and bucked. Compare with the system in figure 219.

Parallel coupling of the partial operations is particularly desirable for harvest of small timber from first and second thinning operations.

Comparisons of performance and costs will show the most advantageous alternatives in a given situation. The analysis can be carried out as demonstrated on page 337-344.

FIGURE 222. A clam harvester in operation. (Skogen, 1983)

Methods of harvesting in mature stands

Harvesting in mature stands provides a wider choice of applicable methods. The conditions favouring mechanization of harvesting in mature stands have been rather easy to cope with due to bigger trees, a higher volume of timber per hectare, and full freedom of movement for the machines in the clearcut areas.

Final harvest operations

For mechanized harvest operations in old stands there are several different harvesters on the market which process the trees into assortments in the stands before a transport of timber by forwarder to the truck roads.

To facilitate the operator's work, automation of some functions of the harvester can be expected.

However, in the beginning of the 1980's the most commonly used mechanized system for tree harvesting in old stands consists of mechanized or motor-manual felling and mechanized delimbing, scaling and bucking. This system may be considered as preceding the introduction of harvesters.

The harvesters are judged to assume the harvesting work to an increasing extent in the 1980's and into the 1990's.

Final harvest operations are partly carried out motor-manually by means of chain saws as in thinning operations. The needs for further technical development of the motor-manual systems primarily concern improvements of the chain saw with respect to vibration, noise and emissions.

Mechanization of bunching is of interest. How can the techniques with long crane booms on forwarders be developed? The main problem is to achieve sufficient reach without an excessive increase in weight and with retained stability of the forwarder.

At harvesting of big timber the motor-manual system with chain saws may be the only alternative because of the size and weight of the timber. Harvesting of trees larger than 80 cm in dbh, 30 m - 40 m in total height and a volume of 4 m^3 or more, which may require careful delimbing and bucking, is normally done by motor-manual systems e.g. in forests with big trees in southern Europe and western United States and Canada.

The assortment (shortwood) method

In Sweden the assortment method dominates harvesting in mature stands as well. Experiences gained from mechanized harvesting systems appear to lead to a shift from the tree methods to the assortment methods. The reasons for this shift may be the greater organizational problems of the tree methods. Large clearcut areas and a continuous removal of timber from the landings are required. There are also problems with the accumulation of limbs and the sensitivity of the tree methods to disturbances in the machine operation.

An increase in the use of mechanized assortment methods is also due to the development of harvesting machines that can operate in the stands. The machines can be used for all processing operations near the stump prior to a direct transport of the timber to the industries. This will obviate the need for costly timber handling in the interfaces of the transport operation.

The rapid development of competitive assortment methods is primarily due to generally favourable terrain conditions. Where the terrain conditions are more difficult with adverse slopes, the tree length methods and the tree methods are preferable as is the case in Norway, the southern parts of Europe with their alpine terrain conditions and in western North America. In Finland development is similar to that in Sweden (Wibstad, 1983).

The assortment method So 1 still is common at harvesting in mature stands, see remark above. Of the entirely mechanized assortment methods So 3 and So 5 were most common in 1983 in most countries (Figure 219). The economics of So 2 is influenced by average size of the timber, more than that of So 4. The method So 2, therefore, is preferable in big timber from harvesting in mature stands while So 4 is more suitable for harvesting operations in small timber. Method So 3 has, with the mechanized felling, replaced So 2 in the last years.

Assortment method (example)

Planned final harvest operation with motor-manual felling

1. Width of parcel may be 15 m, of which the strip road is at least 6 m. The storage zones on each side of the strip road are 3 m wide.
2. The parcel for felling is laid out from the margin of the stand or from an opening and straight into the stand that is to be harvested.
3. The strip roads and the storage zones along the roads are cleared gradually. Bushes and undergrowth are removed. All limbs are put in the strip roads.
4. First, fell trees in each storage zone parallel to the strip road so they can be used as a working bench - one tree crossed over the road.

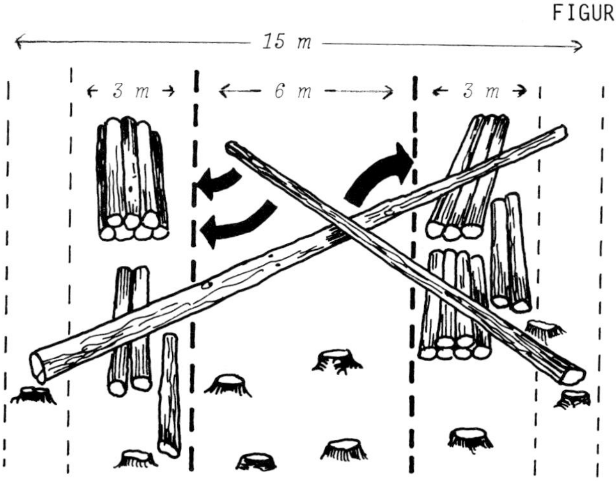

FIGURE 223. Felling of side trees:
- side trees are felled last
- use the piled timber as working benches
- felling is done as for other trees.

5. Use the working bench for the nearest group of trees so that the limbs are laid in the road and the timber can be easily rolled into the storage zones for bucking and high piling.
6. Last, fell side trees standing outside the storage zones and use the piled timber as working benches (Figure 223).

The tree length (trunk) method

The tree length method is used primarily for large areas of harvesting in mature stands with a high volume of timber per hectare and where the trees are big e.g. an average of 0.30 m^3 per tree, requiring great care in scaling of the sawtimber. The tree length method is considered to be slightly less sensitive to the terrain conditions than the assortment method.

The tree length methods St 1 and St 2 are the most commonly applied methods today with a strong trend in favour of St 2 if delimbing machines are used. The methods St 3 and St 4 may be alternatives for future use. St 3 appears to be more suitable for big timber and difficult terrain conditions than method St 4.

Whether bucking should be done at upper or lower landings depends i.a. on the distances of transport and on the conditions prevailing at the upper landing. In most cases bucking at the lower landings close to the industries is preferable.

1. Width of parcel 15 m - 20 m, width of road 4 m.
2. The trees are felled so as to facilitate coupling (choking) and winching
3. Trees felled previously should be used as a working bench of height suitable for delimbing and topping work.
4. Branches and limbs must not cover the timber at the points of coupling (choking)
   - continue according to the same pattern for the remaining trees i.e. utilize the previous trees as a working bench
   - limbs (and branches) must not cover the timber at the point of coupling (the top end)
   - small trees are felled so that two or more trunks can be winched in the same coupling.

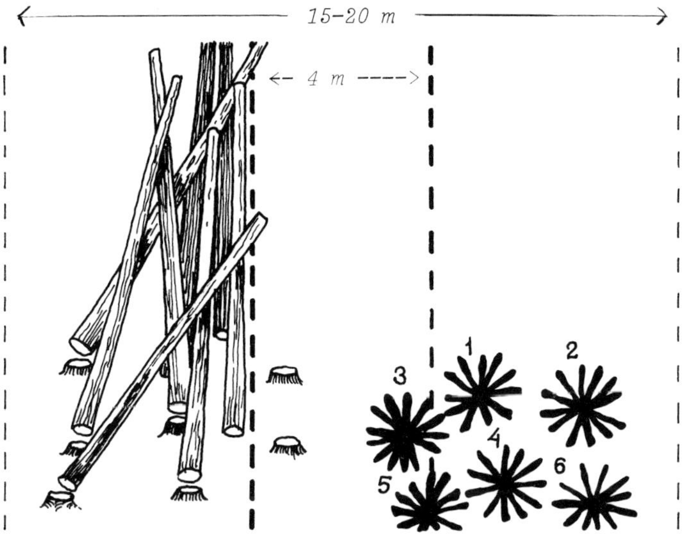

FIGURE 224. The tree length (trunk) method. (Husqvarna skogsteknik, 1982).

The tree method

The tree method is primarily applied in difficult terrain for very limby trees at high elevations, difficult snow conditions, large clearcut areas, average sizes of trees and short forwarding distances.

A big semi-stationary estblishment for delimbing and bucking requires a large quantity of timber for good processing economy.

Since the tree method has a lower transport output (approximately 20 percent) than the tree length method because of the weight and volume of the limbs, it is sensitive to transport distances exceeding 200-300 m in terrain.

The tree method Tr 1 has now been largely replaced with the methods Tr 2 and Tr 3 of which Tr 3 shows the largest increase.

Degree of mechanization

Entirely mechanized forms of harvesting were estimated in 1972 to account for approximately 10 - 30 percent of the total volume of timber harvested by the major forest owners. The extent is expected to increase up to approximately 70 percent in the 1980's for major industries.

An important matter is the establishment of an optimum rate of mechanization that might be required to compensate for the cost development and to deliver timber and manufactured goods at prices competitive on the world market. The rate of mechanization must be geared to the problems of labour supply and social development that will occur as a result of the reduced employment opportunities in the forests.

Mechanized systems with processing in the forests

In these systems tree harvesting work may be entirely or partly mechanized depending on whether felling is done by means of felling heads or chain saws. Delimbing and bucking are carried out mechanically in processors or harvesters.

Mechanized systems with processing at terminal or at industry

Industrial timber handling techniques can be used when the trees are processed in a central place. In addition to delimbing, scaling, bucking and sorting of timber, debarking and chipping can also be done.

Machines and components for such systems are developed. However, it is less well known how the components should be combined in order to achieve an optimum production. The level of production achieved by techniques presently available has indicated that processing terminals generally do not yet produce as expected.

Mechanized systems with limited crews

Work carried out in the stands at tree harvesting can hardly be mechanized without the use of crews e.g. for the existing harvesters or for the future complete tree harvesting machines. However, there appears to be certain possibilities to develop automatic processing of trees or trunks at central landings. Such a development, however, will be rather expensive, at least initially, and it may not be applicable by year 2000.

After intensified development efforts in the 1970's the harvester is expected to become the dominant type of machine used for tree harvesting operations in the 1980's.

Assuming that the forest road net will be further densified in order to reduce the costs of timber transport from the stumps to the industries, we may anticipate that the harvester will be designed further for load carrying functions. Such a development would lead to _a complete tree harvesting machine_ for all work in the assortment method by year 2000.

For a successful application of increasingly complicated operations and technical means in forestry, an efficient training of personnel at all levels is an absolute prerequisite. This cannot be overly stressed. At the same rate as the progress of mechanization, therefore, training activities have been intensified (Cf. Stergiades et al, 1981).

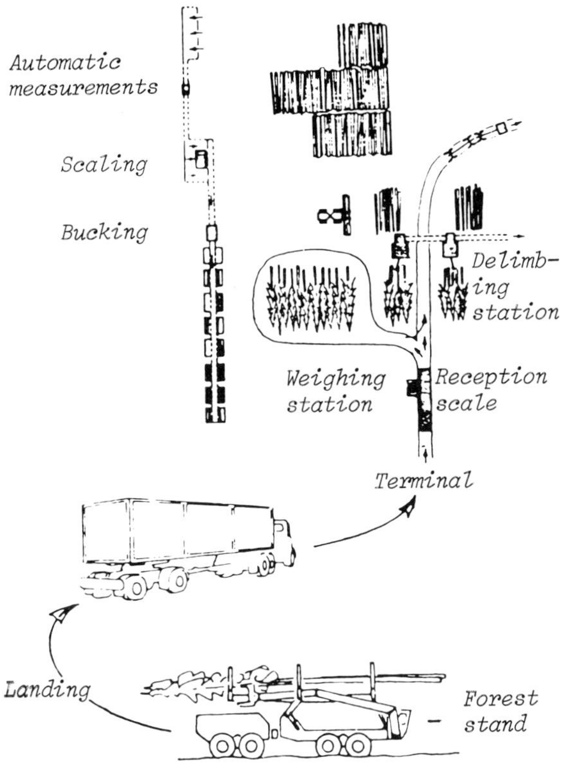

FIGURE 225. Harvesting system for trees with processing at terminals. (Logging Research Foundation, 1983).

The following graph gives an approximate picture of the gradual mechanization of tree harvesting and transport between 1930 and 1990 (Skogsarbeten, 1983).

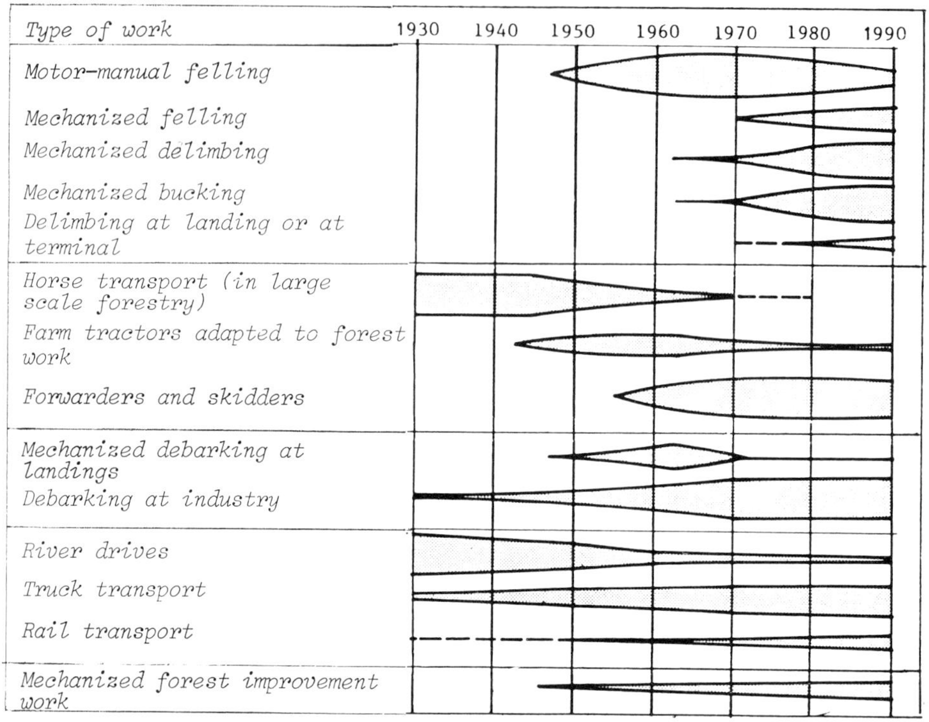

FIGURE 226. Development and trends in the mechanization of forestry work between 1930 and 1990 (Sweden).

Degree of mechanization

Mechanization of thinning operations has advanced slower than expected. Because of the relatively small trees, low volume of timber removed per unit of area, and density of the stands that are in a delicate condition, the main problem has been to achieve a sufficient production. The degree of mechanization of tree harvesting in thinning operations (in Sweden) at the end of the 1970's was barely 15 percent.

In contrast, mechanization of tree harvesting operations in old (mature) stands and transport in terrain has progressed quite rapidly, in many countries increasing from an approximate average of 5 percent in 1970 to 65 percent in 1980.

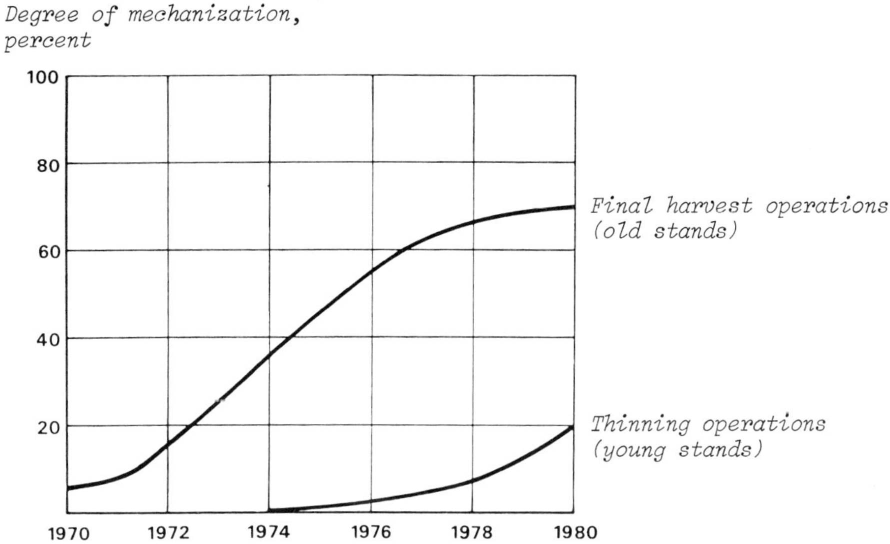

FIGURE 227. Degree of mechanization (Andersson, 1982).

Machine development

The costs of developing equipment for tree harvesting varies strongly with the size and complexity of the machines. Cost of the prototype is often a minor part of the total cost of development. For heavy machines the cost of the prototype is usually less than 10 percent of the total cost.

The cost of developing a processor distributed by various stages is given as an example below. Work on the first stages started in the beginning of 1969 and the project was finished in the middle of 1971. The total cost amounted to 7 mill. SEK (1 $U.S. = 8 SEK (1984). The course of development is shown in Figure 228.

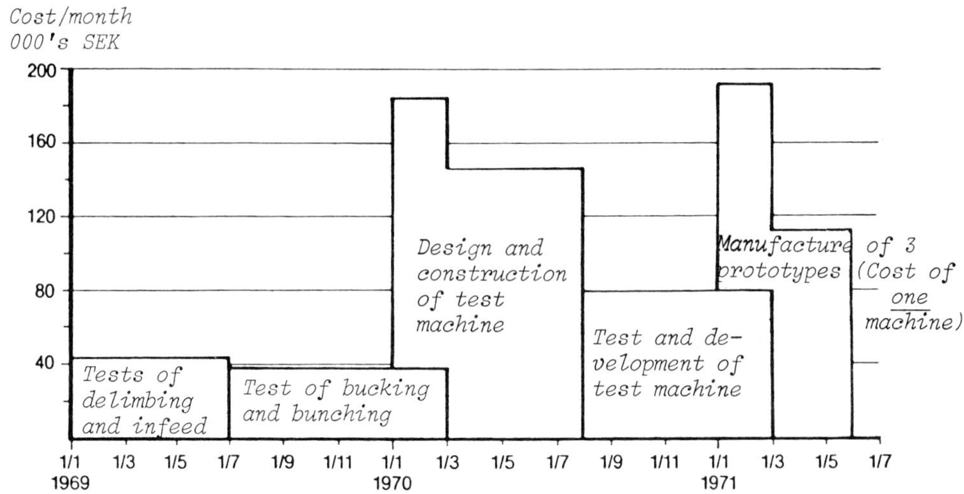

FIGURE 228. Course of developing a processor.

The graph in Figure 228 shows the extent and expenses of a development process for a tree harvesting machine (STU, 1983).

The following table shows how the use of various types of machines has fluctuated during the intensive period of mechanization in the 1970's. The table is based on inquiries with retailers of forestry machines in Sweden.

Table 12. Various types of machines.

Type of machine	1970	1971	1972	1973	1974	1975	1976	1977	1978	1979	Total
Feller	1	3	3	3	6	124	96	25	11	15	287
Feller-skidder	21	37	48	24	0	0	0	0	0	0	130
Delimber	32	47	35	22	32	14	5	9	4	4	204
Delimber-bucker	15	45	56	84	148	219	94	52	98	163	974
Harvester							6	20	17	22	65
Bucking units	16	22	16	18	3	1	0	0	0	0	76
Clam bunk skidder	20	43	19	18	10	3	0	1	0	4	118
Winch skidder	145	27	20	21	17	34	22	13	4	0	303
Forwarder	752	905	762	822	770	606	496	450	352	387	6302

Potential man-machine systems for thinning

Mechanization of thinning appears necessary, the conventional motor-manual form of thinning probably not being capable of developing towards any essentially higher output.

Thinning is a stand treatment with a dual purpose viz. an accelerated production of a crop of timber and improvement of the future value of the remaining stand (Jevins et al, 1976, Arvidsson et al, 1983).

Comprehensive experiments are carried out in order to find profitable forms of entirely mechanized thinning, primarily in young stands.

Given below is a presentation of new techniques in thinning which have been subject to research and trials. Three different man-machine systems have been studied by theoretical performance analyses combined with simultaneous experiments and studies in test bench and in the field.

FIGURE 229. A potential machine for harvesting in swaths.

Thinning by means of a machine for harvesting in swaths

One system of thinning has been the so-called corridor thinning, i.e. clearing of aisles or corridors in the stands by means of machines designed for harvesting in swaths. The principle design of the analyzed machine for harvesting in swaths is shown in Figure 229. Since this machine carries out a geometric thinning, the valuable biologic effect of a conventional thinning is lost. In conventional thinning the most vigorous and biggest trees are left while the weak and small trees are removed.

Thinning by means of a tower crane

A second machine called tower crane could be used for harvesting along strip roads and for thinning relatively far out from the strip road. The trees would be processed vertically and above the crown canopy (Sundberg, 1970).

The felling mechanism of the crane would be operated over the shortest possible distance between the trees which is a great advantage. This feature provides an opportunity to reduce the cycle time per tree to 25 - 30 Cmin (Cmin = one hundreth of a minute), corresponding to a production of 3 - 4 trees per minute (Figure 230).

FIGURE 230. Potential design of a tower crane for thinning.

A development of the delimbing operation and the forwarding of timber along the crane boom down to the cradle should not meet with any major technical difficulties. However, there may be some potential problems with stability, primarily during the movement of the crane from one position to the next.

Thinning by means of multi-tree fellers

A third type of machines designed for thinning is a base machine with multi-tree felling mechanism. The machine is equipped with a hydraulic crane and a felling head which can buck and recover 2 - 3 trees (perhaps more) from the stand in each cycle (Bredberg & Moberg, 1972).

The total effect of thinning is a criterion on the value of the man-machine system. Such a criterion is obtained according to the following principles of calculation (Figure 231).

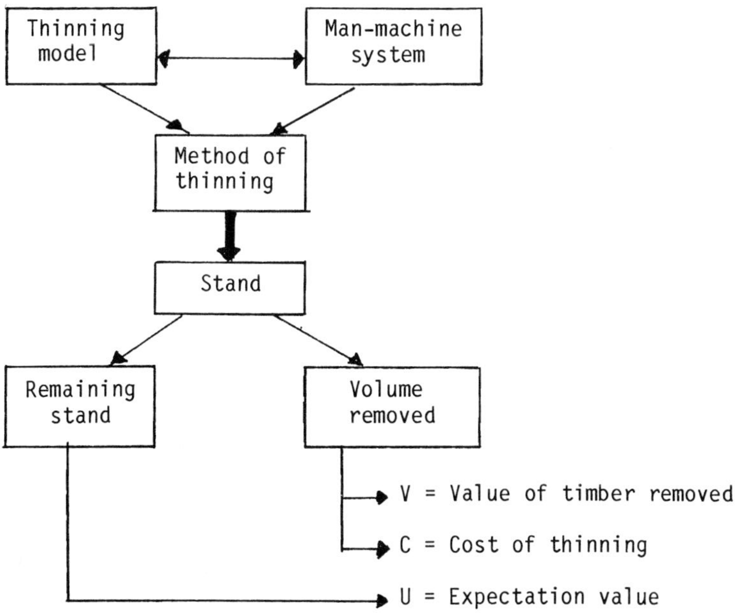

FIGURE 231. Principle of calculating the economic result of thinning by means of different machine systems.

Criterion for effect of thinning (E):

$$E = V - C + U \quad \text{SEK/ha}$$

As an example of conditions in the northern coniferous region a comparison between the economic results of thinning by means of three different machine systems has been based on calculations on the criterion for effect of thinning presented above. The comparison indicated that the highest net return was obtained for dense corridor thinning by means of a machine for harvesting in swaths. When the expectation value is included, thinning by means of the multi-tree machine was most advantageous. On better sites (average height of dominant trees 28 m at age 100 years) thinning by the tower crane produced a good economic result.

It was also shown that:

1. Expectation value is strongly influenced by the rate of interest and by the method of thinning.
2. Geometric (corridor) thinning gives the best effect only at a high rate of interest, approximately 10 percent and higher.
3. At 5 percent rate of interest a heavy selective thinning gives a high return which should be compared with the considerably lower return obtained for the untreated stand.
4. The advantage of using a machine system that can be operated for a biologically proper thinning was obvious from the comparisons.

The problem encountered at a comparison of this kind is the evaluation of the future net return from the stand. In practice a forest owner usually does not thin when the cost of thinning exceeds the sales value of the timber removed. The future gains obtained by thinning appear remote and uncertain. The forest owner, therefore, may consider a costly thinning as a form of long term investment.

It would be valuable for the forest owner if a machine for thinning could be developed with such efficiency that the cost of harvesting and the value of timber are at least equal.

It must also be considered of great interest that research and development in forest technology be promoted so effectively that the problems of thinning can be tackled and solved as soon as possible (Herpay, 1981).

Integration of harvesting and transport

Harvesting and transport in terrain can be coordinated or integrated to a large extent by means of modern technology. Integration primarily saves time and time is big money when expensive machines are being used (Cf Grammel, 1975 and Mihac, 1975).

Performance analysis of a machine designed for thinning

The following chapter gives a performance analysis of a potential thinning machine with a high integration of the partial operations and transport. Integration is assumed to be so high that time required for transport between the felled trees does not have to be presented separately in the time formula for the machine (Cf. Bol, 1978). (Hedbring et al, 1968).

The performance analysis is also an example of a method of analysis for the evaluation of a man-machine system. The potential machine, a harvester-carrier or, preferably, a logger in a project named the Garpprocessor, has been subject to a comparison with other Man-machine systems within a project called Entirely Mechanized First Thinning (HMG).

The performance analysis shows the sensitivity of production with respect to variations in the size of timber. Thus, production is doubled when the average tree size increases from 0.05 m^3 to 0.14 m^3.

A comparison carried out for a machine with multi-tree feller would probably indicate increased production and less dependence on variations of tree size.

The difference in performance between the alternatives shows how much production can be increased if transport can be entirely integrated into the harvesting system. A production increase of approximately 30 percent is obtained in alternative I in which the machine operates while moving along the strip road. This confirms a principle stated previously in the mechanization thesis (page 209).

Analysis of a man–machine system for thinning

HMG 8 logging machine

Description of machine

The machine is built on a four-wheel drive chassi for articulated steering. Total width is 3.0 m and length is 8.0 m. The machine is operated by means of a hydrostatic transmission. Engine power is 95 kW and machine weight is 15 tons. Felling is done by means of hydraulic shears mounted on a telescopic boom with a reach of 14 m. Pneumatically operated and curved knives are designed for delimbing and feeding is done by means of rollers.

Timber is bucked into standard length, 3 m, and hydraulic shears carry out topping (Staaf, 1972).

The operator's cabin with delimbing and bucking mechanisms is placed on a turntable with centre of gravity above the front axle of the machine.

A revolving telescopic boom is mounted on top of the operator's cabin above the midpoint of the front axle.

On the rear carrier of the machine is built a timber cradle with a capacity for 7.5 m^3, or alternatively, a bunch of 1.5 m^3. The timber cradle, the side stakes of which are jointed, can be tipped sideways by a hydraulic mechanism.

The combined delimbing and bucking mechanisms are automatically aligned with the telescopic boom in preparation for processing. After the tree has been taken, the processing mechanisms operate in a fixed position relative to the machine (Figure 232).

Crew: 1 man

Description of the method

The machine is moving along a road for thinning. From each position the road section in front is cleared and the stand within a sector of maximum 45° on each side of the road is thinned. Good stability is achieved since felling is done in front of the machine.

FIGURE 232. A potential logging machine (HMG 8).

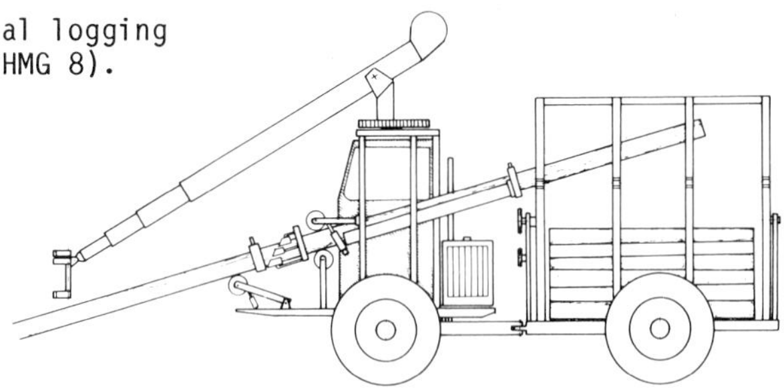

FIGURE 233. Principle of thinning procedure for the logging machine HMG-8.

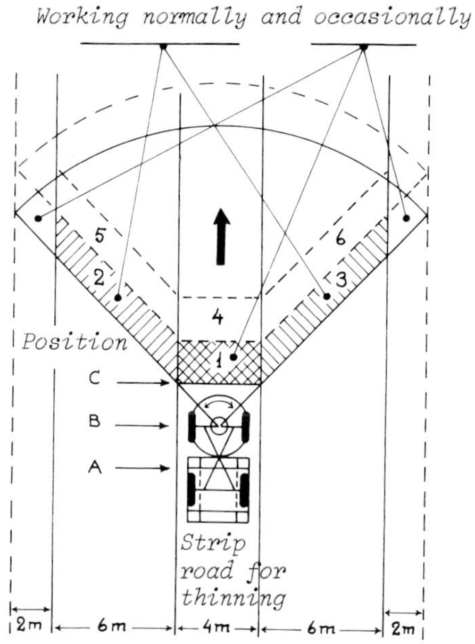

The trees are severed from the stumps, hauled in horizontal position to the machine and fed into the processing components for delimbing, bucking and topping. All residues, composed of limbs and tops, are collected in front of the machine into a layer carrying the machine and protecting the ground along the road.

Timber is bucked into 3 m length, the logs being collected in the rear cradle. A certain degree of positioning of the rear carrier may be necessary for the collection, while the folding side stakes can serve as guiding braces for logs that are fed at an oblique angle to the carrier.

While the tree is being processed, alternatively while the machine proceeds clearing the road in front at a slow speed, a new tree is hauled in by means of the telescopic boom. When the first tree is finished, the processing components are turned into the direction of the crane and the next tree is fed in directly.

While the machine is moving slowly from position A to position B, road section 1 is cleared. From position B the sectors 2 to the left and 3 to the right are thinned (Figure 233).

During the next slow move from position B to position C, road section 4 is cleared.

From position C the sectors 5 to the left and 6 to the right are thinned.

The machine can operate in all other parts of the 90° sector within a radius of 14 m.

When the cradle is filled the timber is tipped at a collector road or an access road for further transport by truck.

Example of performance

1. Moving between the positions

1.1 Moving while clearing the road for thinning

$$T_{1_{I}} = 0$$

1.2 Alternatively separate movement between the positions

$$T_{1_{II}} = \frac{10\ 000}{32} \times \frac{G}{U} \left(\frac{K \times 2}{V} + 8 \right) \text{ cmin per tree}$$

G = volume of average tree, m^3

U = volume of timber per hectare, m^3

K = allowance for winding

V = speed of travel, m/cmin

2. Movements of crane

2.1 Turning into position for felling (45° left and return , 45° right and return) corresponding to 180°, or 20° per cmin for all trees per position

$$= \frac{10\ 000}{32} \times \frac{G}{U} \times 9 \text{ cmin per tree}$$

Crane moving toward tree

Felling 20 m according to pattern 1-2-3 6.4 m or 0.45 m per cmin = 14 cmin per tree.

Felling 16 m acording to pattern 1-2-3 5.3 m or 0.45 m per cmin = 12 cmin per tree

2.3 Positioning, average time, 10 cmin/tree
Time differentiated for 0.05 m^3, 0.10 m^3
and 0.14 m^3 per tree 9.0 10.0 11.0 "

2.4 Shearing at stump height, average time 7 cmin/tree
Time differentiated for 0.05 m^3, 0.10 m^3 and
0.14 m^3 per tree 6.0 7.0 8.0 "

2.5 Hauling in tree, tree falling
Felling 20 m, 6.4 m 17 cmin/tree
Felling 16 m, 5.3 m 15 cmin/tree
Time differentiated for 0.05 m^3, 0.10 m^3 and
0.14 m^3 per tree 13.0 15.0 17.0 "

2.6 Putting the tree into processor 5 cmin/tree
Felling 16 m, time differentiated for
0.05 m^3, 0.10 m^3 and 0.14 m^3 per tree "

$$T_{2_I} = \frac{10\ 000}{30} \times \frac{G}{U} \times 9 + 12 + (9.0\ \ 10.0 \text{ or } 11.0) + (6.0\ 7.0 \text{ or } 8.0)$$
$$+ (13.0\ 15.0 \text{ or } 17.0) + 5 \text{ cmin per tree}$$

Felling 20 m $T_{2_{II}} = \frac{10\ 000}{32} \times \frac{G}{U} \times 9 + 14 + 10 + 7 + 5$ cmin per tree

3. Delimbing, bucking

These partial operations are done when crane is operating
$T_3 = 0$

4. Empty timber cradle

Tipping of timber cradle, 45 cmin for a bunch of 1.5 m^3
200 cmin for a load of 7.5 m^3

$$T_4 = \frac{\frac{45}{1.5}}{G}$$

$T_4 = 30\ G$

Time formula for feller-delimber-bucker

Alternative 1. $T_2 + T_4$

Alternative 2. $T_1 + T_2 + T_4$

Example of performance

Conditions

	A	B	C
G	0.05 m^3 per tree	0.10 m^3 per tree	0.14 m^3 per tree
U	50 m^3	50 m^3	50 m^3
K	1.4 allowance[1)]	1.4 allowance	1.4 allowance
V	0.15 m per cmin	0.15 m per cmin	0.15 m per cmin
L	8 m per tree	10 m per tree	11 m per tree

1) allowance for distance of winding road.

Time formula for harvester-carrier

Alternative I. Felling in 16 m wide swath, not separate movement between the positions, differentiated time for felling

$T_{2_I} + T_4$

Alternative II. Felling in 20 m wide swath, separate movement between the positions, not differentiated time for felling.

$T_{1_{II}} + T_{2_{II}} + T_4$

Results

Alternative I	A		B		C		
Turning	3.0		6.0		8.0		cmin/tree
Crane to tree (5.3 m)	12.0		12.0		12.0		"
Positioning	9.0		10.0		11.0		"
Shearing	6.0		7.0		8.0		"
Hauling in tree	13.0		15.0		17.0		"
Infeed	5.0	48.0	5.0	55.0	5.0	61.0	"
T_4	1.5		3.0		4.5		"
Total time Alt. I	50.0		58.0		66.0		"
	10.0		5.8		4.7		min/m^3
	6.0		10.3		12.8		m^3/h
	30.0		51.5		68.0		m^3/shift
	763		763		763		SEK/shift
	25:40		14:80		11:92		SEK/m^3
Alternative II							
T_1II	8.4		16.8		25.2		cmin/tree
T_2II	55.5		57.9		59.6		"
T_4	1.5		3.0		4.5		"
Total time Alt. II	65.4		77.7		89.3		"
	13.0		7.7		6.0		min/m^3
	4.6		7.8		10.0		m^3/h
	23.0		39.0		50.0		m^3/shift
	763		763		763		SEK/shift
	33:20		19:60		15:30		SEK/m^3

Comments

The machine HMG 8 can also operate as a carrier with a timber cradle holding approximately 10 m^3 (volume of piled timber). To facilitate comparisons with other machines the alternative presented has been based on a load size of 1.5 m^3 of solid wood.

When transport distance is short, the machine, if equipped with a bigger cradle, should be compared with a processing machine with separate carrier.

Work studies

Work studies as a source of reference

Work study activities, which were initiated on a modest scale in forest operations about 1920, did not primarily deal with matters of rationalization but were intended to produce a basis for collective work and wage agreements Gradually, however, rationalization has become an increasingly important objective of work studies (Luthman et al, 1942).

The following concepts of ergonomics, work studies and work have been defined for the Nordic countries in the publication "Nomenclature of Forest Work Studies" prepared by the Nordic Council of Forest Work Studies (NSR).

Ergonomics

Ergonomis is the science of work and its productivity, its share in Society's total result of production, and ways of measuring productivity. Objective of ergonomics is the furtherance of knowledge on:

- Work
- Men at work
- Machines, tools and other equipment used as means of work
- Interaction betwen these elements and their optimum coordination

Work studies

Work studies are one of the most important sources of reference in ergonomics.

- Work studies are systematic investigations of work, men at work and the- technical conditions carried out for the purpose of gaining knowledge.

  The fields in which work studies are primarily applied are:

  - Rationalization, which is a conscientious, systematically organized activity, aimed at improving work output in a given field of activity.

- Pricing of work on the objective basis of work studies.

Work

Work is Society's active, original factor of production in its direct (manual) and indirect (capital induced) forms. From a study point of view, work is to be perceived as an active occupation aimed at changing the form, location or condition of the object of work. Work can be carried out physically, mentally or in some other way of participation in a process.

Objectives and means of work studies

Current concepts of objetives and meaning of work studies may be summarized in the following way:

Work studies are means of raising productivity in a given field of activity by a conscientiously high utilization of the available resources, such as establishment, labour and material (Cf. Popa, 1979).

The mental and physical capabilities of Man have also been increasingly taken into account at the design and changes of production systems. Particular attention has been paid to the limits of physical and mental stress, health hazards, risks of accidents etc.

Thus, work studies are primarily used for technical - economic rationalization. Work studies are also an important means of improving work safety and health care.

Various forms of work studies

The choice of work study method most suitable for a given investigation is influenced by i.a. the object and purpose of the study.

Object of the study

The object of study may be a person, a crew, or a machine which gives rise to

- Studies of persons
- Studies of crews
- Studies of machines

Purposes of study

The purposes of a work study may be to study a method, performance values for negotiated agreements, or to serve as a basis of calculations and analyses. The following types of studies, therefore, can be distinguished:

- Studies of methods
- Studies of agreements
- Studies of calculations

Methods of study

The methods of study that may be considered are i.a.

- C-min studies
- Frequency studies
- Statistics

Objects, purposes and methods of studies mentioned above together constitute 3 x 3 x 3 = 27 different layouts.

Each layout has its advantages and disadvantages. At the choice of study form, which is to be considered as a method of measuring a certain work output, it is important to clarify carefully the objectives of the work study.

Measurements

What is to be measured? Measurements concern time required, spacial changes, energy required or wear of machines, mental or physical stress.

Time studies

Time required can be measured directly by special time studies, usually in the form of c-min (abbreviation of centi-minute) studies, which are recorded to the nearest one hundredth of a minute. Time is recorded when work begins, when it changes nature i.e. at transition from one work process to another, and when work is finished.

The study can be carried out either by recording the time of each work process by setting the stop-watch at zero, the so-called zero method, or by recording the time elapsed from the original start, the partial time of each process being obtained by subtraction of the current readings, the so-called continuity method.

Time required can also be measured roughly be means of an ordinary watch or by means of vibration clocks mounted on machines e.g. tractors. Movies and tape recorders can also be used in special cases, e.g. a movie of entire crews.

The c-min study has its advantages when an elaborate basis is required e.g. for development of a method, and it is most suitable for studies of a person or a machine. However, it is an expensive method.

Frequency studies

Frequency studies are based on probability analyses and they are used for determination of the relative proportions of various work phases. This method is based on recordings or measurements at random or regular intervals of time only. The phase that is current at the moment of recording is observed. If a sufficient number of observations are made, the probable distribution of work by various phases is obtained.

The result of a frequency study can also be used as an approximate estimate of the absolute time requirements. The regular time intervals are normally about 1/4 - 1 min. However, 5-minute intervals occur as well.

The advantages of frequency studies are associated with i.a. the possibility to study several workers or machines simultaneously. This type of studies is relatively inexpensive to carry out.

Studies of statistics

Studies of statistics record the work phases which are dominant during the period when statistics are being collected. In an almost nationwide collection of statistics on time input and earnings in the forest operations, current work is recorded to the nearest five minutes.

Studies of statistics are based on the collection of data on time input and performance over long periods. These studies are extremely short on detail and they can be carried out by the workers themselves (Staaf, 1953).

Application of work studies

The layout of a work study is usually preceded by a minor study, sometimes called preparatory study or pilot study.

Before the study is initiated, the observer should be thoroughly acquainted with the current elements of time.

Elements of time

Figure 234 shows the place of time elements in a scheme that was established by NSR in 1963.

Definitions of various elements of time have been presented in "Nomenclature of Forest Work Studies" (NSR) and in Dictionary of Forest Terminology (TNC 71-1978).

Deviations from the scheme may occur i.a. in "Time Elements in Machine Operations" published by Logging Research Foundation in 1969 (See Dictionary of Forest Terminology 1978).

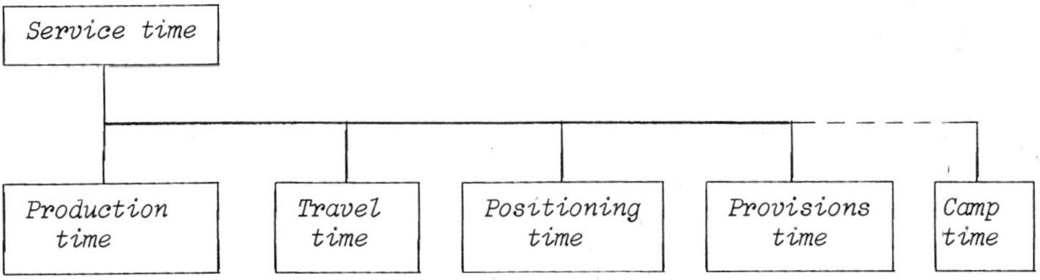

A more strict break-down of production time has been made in the following scheme:

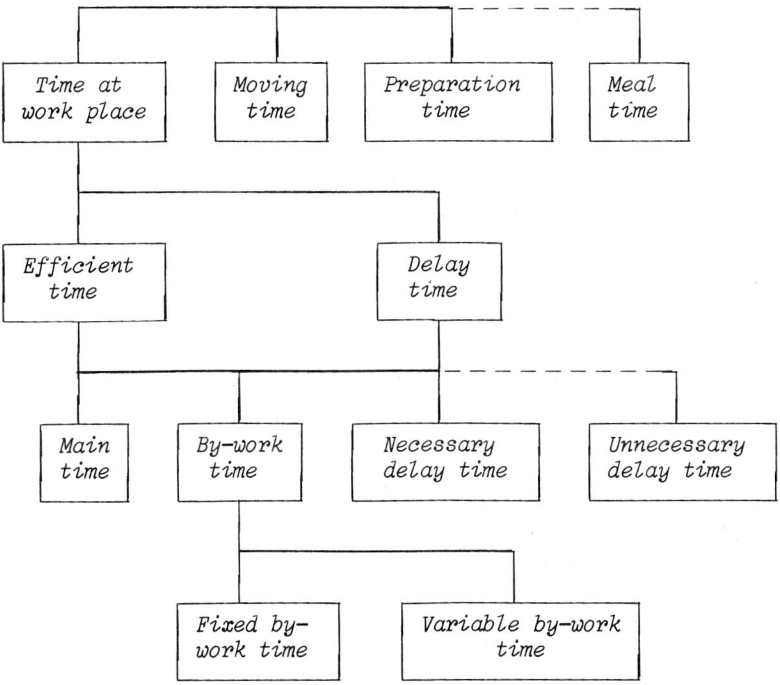

FIGURE 234. Elements of time applied in forest operations (NSR, 1978).

Purpose of work studies

From the beginning the purposes of work studies in forest operations were to:

1. Clarify the influence of various forest conditions on work difficulties
2. Study and select the most feasible technical means
3. Record work performance

When work studies were initiated and organized in forest operations on a large scale in the latter part of the 1930's, the objective was to establish a basis for equitable collective agreements on work and wages primarily in felling, processing and horse operations.

Studies of rationalization

Simultaneously with the pure time studies, it is also desirable to improve the conventional methods of work by rationalization.

The technical evolution has brought about a greater interest in rationalization, introduction of new means and methods of work. Studies of methods were initiated and they are now of dominant importance at studies of forest operations (Embertsén, 1976).

When a new harvesting system is to be tried in today's situation, it is recognized that the organizational layout of work can be considered to be a distribution problem involving people and machines used in the system. Optimum combinations of all the functions carried out by people and machines are explored by means of various studies of time and methods e.g. certain combinations of c-min studies and frequency studies. For instance, when a new man-machine system is to be formulated, the following steps can be followed:

1. The purpose of the man-machine system is identified by means of a description of objectives.
2. Description of all necessary functions
3. Distribution of functions by manual and mechanized operations
4. Further descriptions of all work functions i.e. all activities that are required in order to carry out a function or a group of functions, and auxiliary means required
5. Specifications of labour requirements, giving the need for knowledge, ability, skill and personal traits required for each work task.

6. Positions or services are classified including assignments and areas of responsibilities
7. Development of components for the system, e.g. machines are designed, manufactured and installed, personnel is selected, educated and trained for its tasks.
8. Components of partial systems are coordinated. The partial systems are then composed into a complete man-machine system.

As an example of a relatively common application of forest work studies may be mentioned studies of methods for a central place of processing arranged to achieve the best possible organization and work conditions for a crew with a number of machines. It is also possible today to carry on studies for crews by means of so-called check lists and forms in order to facilitate a current day-by-day rationalization of the forest operations.

Forms of work studies in forest operations and in industries

The methods used for pricing of work in forest operations have been different than the methods used for time studies and pricing of work in industries (Hilf, 1957).

Conventional tree harvesting contains for each tree a number of partial operations. A calculation of the agreed piece rate for each individual tree according to methods used in industries would require an unreasonable amount of time study data (Figure 235) (Kilander, 1961).

In view of the large variation in working conditions because of varying sizes, limbiness, terrain etc. such a procedure would be entirely unrealistic. In forest operations it is also highly doubtful whether the time study man is able to evaluate objectively the performance of a worker in relation to that of a theoretical normal worker as is done in industrial time studies.

In forest operations, therefore, the general agreement gives a fixed-piece rate price per tree, volume, length or some other unit of payment for a given harvesting project with uniform conditions. The negotiating parties in forest operations have then used an average output per working day according to comprehensive statistics on time input, earnings and performance as a basis for the establishment of the wage level (Figure 236).

To formulate price lists for conventional narvesting, however, thorough

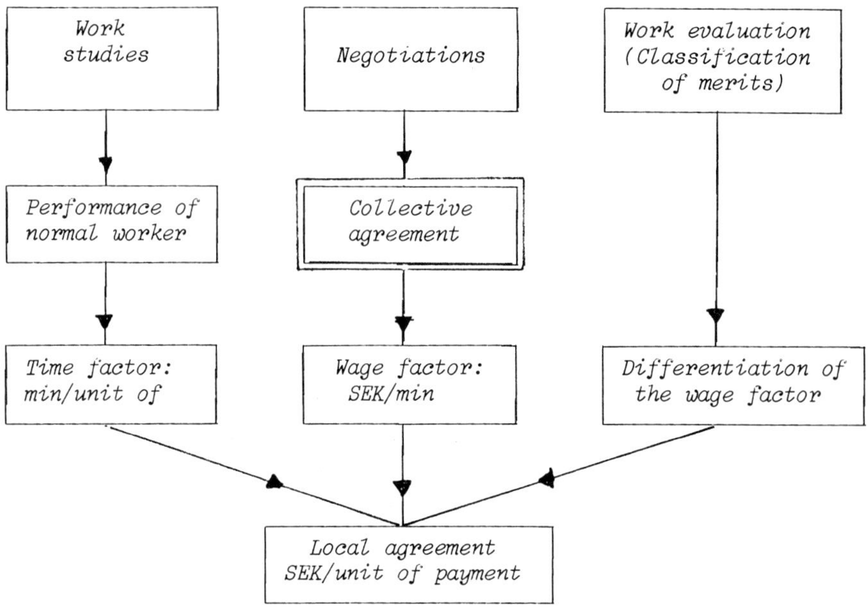

FIGURE 235. System for pricing of work in industries.

knowledge of the relationships between time requirements and sizes, species, degrees of processing, weather, season etc is needed (Mattsson-Mårn, 1956).

Key work of an agreement in forest operations

Difference in _relative_ time requirements at felling by means of a chain saw between a big tree and a small tree is usually rather slight for two different workers. At any rate this difference is considerably smaller than the difference in _absolute_ time requirements which would occur between the two workers. When all relative time requirements in tree harvesting are known, a partial operation can be used as key work.

When the key work is established and a piece rate is set, e.g. felling of a pine tree with a diameter of 20 cm at breast height, known time relationships are used in order to obtain corresponding piece rates, e.g. per tree, for all other diameter classes between for instance 10 cm and 40 cm for spruce, for various densities of felling, various assortments at various minimum top diameter etc.

In the work study forms used in forest operations, therefore, it is not necessary to adjust the time recorded to some theoretical performance of a normal worker.

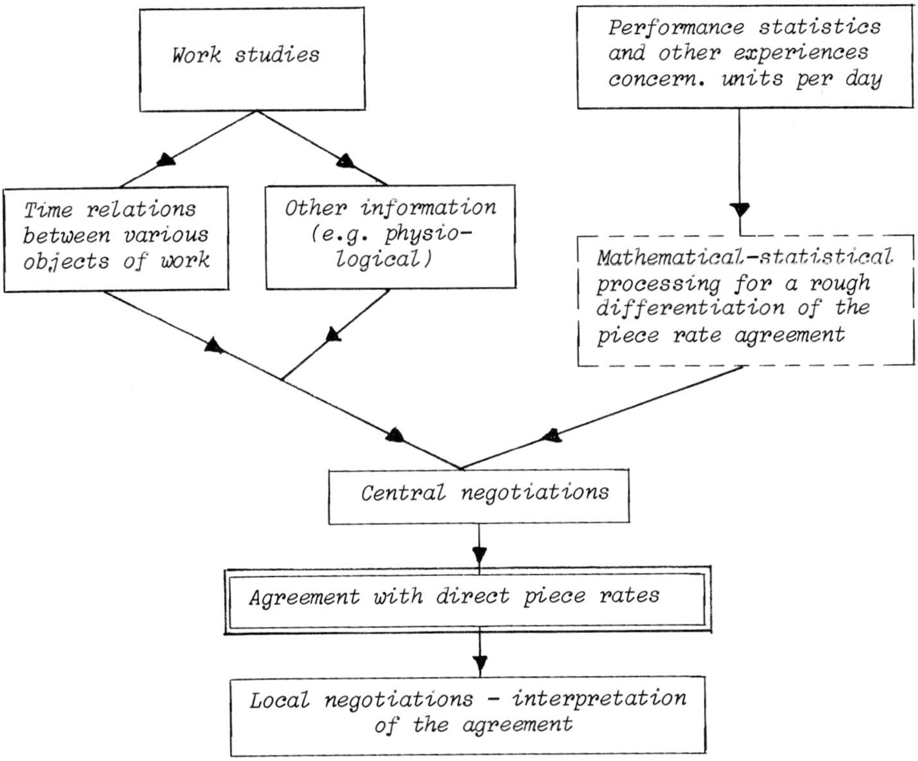

FIGURE 236. System of pricing in forest operations (Kilander, 1961).

Elementary time systems

Other methods of work studies include i.a. basic manual movements. Information on time requirements for closely defined movements, e.g. reaching, moving, turning, etc is obtained. The time data are used for a build-up of the total time required for a whole operation. Combinations of timed basic movements are called elementary time systems of which there are several different kinds. Two systems are used in Sweden, the MTM and the work factor systems. Studies of these systems are applied i.a. in work shops.

Work physiology

Physiological capabilities and limitations of Man

The physiological capabilities and limitations of Man can be described in several different ways (Figure 237).

When attempts are made to place the right persons in the right places, a medical-physiological analysis of a person's work potential may be very valuable (Lundgren, 1964).

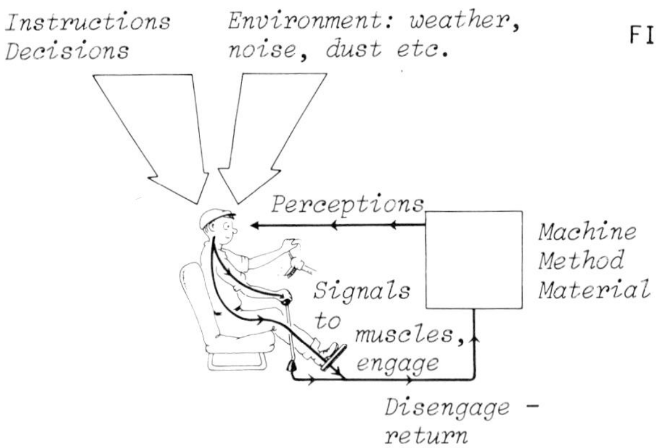

FIGURE 237. Physiological limitations of Man in a work situation.

Check lists

Ergonomic analyses can be carried out by means of check lists in order to explore the working conditions (Table 13).

When a person's limitations are evaluated, it is important to clarify various types of tolerance limits which can be of an individual, medical-physiological or performance nature (Kaldy, 1979).

Individual limitations

The following individual limitations may be distinguished:

- Intolerable zone, which must be avoided
- Discomfort zone, which is rather common in practice, i.a. due to the individual level of endurance and because of ergonomic negligence
- Comfort zone, which is the object of ergonomic research provided it is

feasible from a medical point of view.

Table 13. Guiding values for the evaluation of temperature in a tractor cabin, °C (Logging Research Foundation, 19).

Place of work	Uncomfortably cold	Cold	Comfortable	Warm	Uncomfortably hot
Mostly in cabin	5	5-15	15-22	22-30	30
Both outside and in the cabin (cool season)	0	0- 8	8-15	15-24	24

Medical limits

The following two medical limits can be distinguished:

- Limit beyond which serious injuries occur
- Limit beyond which light symptoms occur difficult to define but warranting intensified research

Physiological limits and performance

The following limits affecting performance may be distinguished:

- Limit beyond which considerable exhaustion occurs
- Limit beyond which performance is affected. This is a physiological limit difficult to establish but very important from a technical point of view.

Physiological and psychological measurements of work

Technical measurements in the form of work studies were discussed in a previous chapter. For a proper evaluation of the work input supplementary physiological measurements of work are necessary. In addition there are the matters of psychological observations and stress caused by e.g. responsibilities and intellectual work.

Physiological measurements

Physiological input or efficiency of Man at work can be calculated by means of measurements of energy turnover per time unit. The measurements can be carried out directly on the basis of oxygen intake, indirectly by pulse rate counts or by measurements of lung capacity.

Energy needed for muscle work can be produced by aerobic and anaerobic processes.

- Energy from unaerobic processes can be supplied immediately to the muscles for short but heavy work performance. For a young, well trained person the output may amount to 1.5 kW for a period of 5 seconds.
- Energy from aerobic processes can be utilized for lengthy work performance. For a young, well trained person output may amount to 0.4 kW for a period of 5 min., or 0.2 kW for a period of one hour. Maximum output depends on the amount of oxygen that can be supplied by the lungs and the circulatory system.

If work load exceeds the maximum output of energy from combustion, additional energy is supplied by the anaerobic processes. Lactic acid is produced and accumulated. Work must soon be discontinued since oxygen deficiency occurs, and oxygen must be replenished after work is finished.

The inhaled amount of oxygen can be used as an indication of the work load (Figure 238).

Approximately 5 kcal is obtained for each litre of oxygen consumed, if energy is produced by combustion.

Exhaled air, the oxygen content of which is measured and compared with that of inhaled air, can be collected in a so-called Douglas bag.

Variation in work capability

Efficiency of work is always the relationship betwen output and input energy. Efficiency is of a practical interest e.g. at the choice of correct method of work or tools for manual work. A choice can be made by comparing work results with the amount of oxygen consumed, pulse rate or quantity of air inhaled.

Work capability of a person at lengthy peformance of work can be measured e.g. on a test bicycle (Figure 239).

If two persons with different maximum oxygen intake carry out the same lengthy work, they will utilize a different proportion of their maximum capability. A trained person will utilize 2/5 of his capability while the untrained person will utilize 4/5 of his capability. In practice, if the persons are equal in other respects, this means that the untrained person must slow down his work rate or stop, while the trained person is able to continue at an unchanged rate.

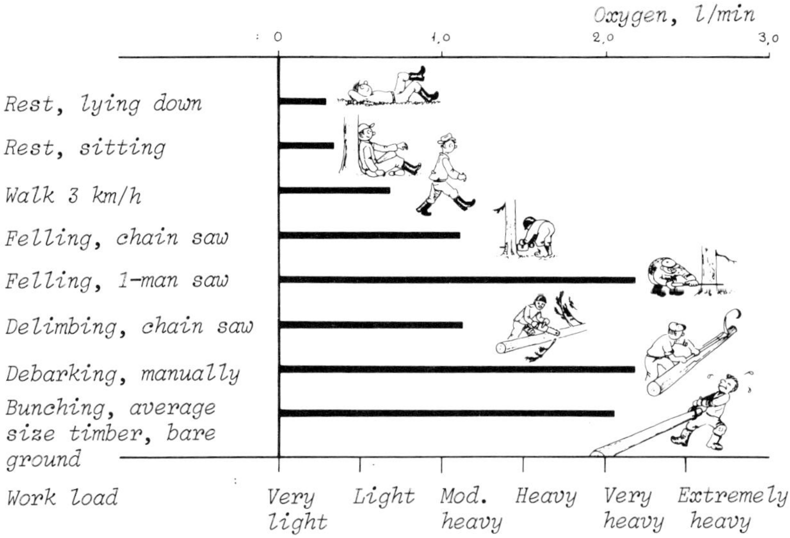

FIGURE 238. Inhaled amount of oxygen is a measure of work load.

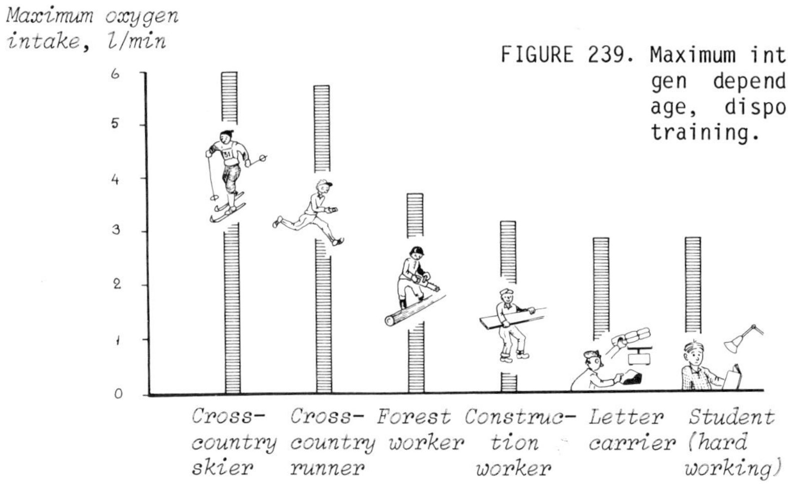

FIGURE 239. Maximum intake of oxygen depends on sex, age, disposition and training.

Physiological work load

The physiological load at a certain type of work is subject to individual variation. Load, therefore, is measured in relative values in contrast to a given output which is expressed in absolute values.

Most common methods of measuring work load:

1. Oxygen intake in certain types of work in relation to the maximum oxygen intake capacity of the individual. If this ratio exceeds 0.5, it can be considered that continuous work is not possible without rests or breaks.
2. Relationship between work requirements and the maximum capability of an individual for exerting working power in a given posture and direction.
3. Lactic acid content of the blood is an expression of strain at heavy work, such as certain athletic performance.
4. Oxygen deficit expressed in litres of O_2
5. Pulse rate in relation to the maximum pulse rate
6. Pulse rate at standard work
7. Body temperature is a measure of work load. It is also influenced by heat stress at the place of work
8. Perspiration is a measure with the same range of usefulness as body temperature
9. Rate of breathing
10. Subjective evaluations by means of psychological interviews, inquiries and standard values.

Need for physiological measurements of work

Need for physiological measurements may occur in many different situations:

1. Physiological measurements of work may be particularly useful for studies of heavy work under hot conditions
2. Physiological measurements of work can be used to decide whether a person is overworked
3. If a sufficiently large group is studied, physiological measurements can give general information on the physical requirements to be applied when recruiting personnel
4. Measurements may occasionally lead to the establishment of a "physiologically normal work rate"
5. Physiological measurements of work may be an aid in demonstrating difficult situations for women, middle-aged men etc.

6. In certain instances measurements may facilitate a more objective evaluation of work, judgement of performance and calculation of the need for breaks.
7. Physiological measurements of work, therefore, are expected to become a more common source of information at certain studies of methods.

Table 14. Oxygen intake capacity, litres/min. for men of various ages. (weight 72 kg) (Åstrand, 1960).

Age	Low	Slightly low	Average	High	Very high
20-29	2.79	2.80-3.09	3.10-3.69	3.70-3.99	4.00
30-39	2.49	2.50-2.79	2.80-3.39	3.40-3.69	3.70
40-49	2.19	2.20-2.49	2.30-3.09	3.10-3.39	3.40
50-59	1.89	1.90-2.19	2.20-2.79	2.80-3.09	3.10
60-69	1.59	1.60-1.9	1.90-2.49	2.50-2.79	2.80

The table values show that the rate of decline in the maximum oxygen intake capacity at increasing age on the average corresponds to approximately 30 percent between age 25 and age 60. It is also realized that the individual variation is very large in each age class (Table 14).

Combinations of work, breaks and rest

It is a rather common practice that forest labour works very intensively during an abbreviated work day with a limited number of breaks and rest intermissions. Such a mode of work is less feasible since physical exhaustion can be caused by extended periods of strain. Breaks and rest, therefore, are needed for physical and mental recovery and for a reduction of health hazards (e.g. noise, vibrations and accidents).

The length of breaks and intermissions may vary. In forest operations it has, therefore, been recommended that the work day be divided into four work periods of 2 hours each separated by three intermissions for meals and coffee, and that breaks of 5 min be taken every hour between the intermissions.

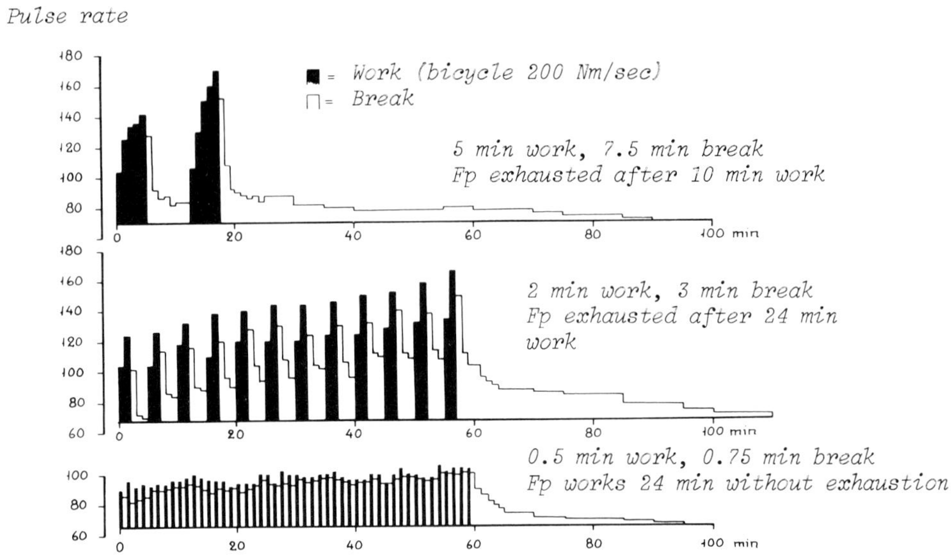

FIGURE 240. Frequent shifting between work and breaks or intermissions will reduce strain (Fp = object of study).

Changes in the mode of work may have a rest effect if strain is reduced and/or other muscles are put to work. In manual forest work, variations in the strain of various operations serve to provide for a reduction of the work load (Figure 240) (Hilf, 1957).

Changes in the mode of work also reduce the strain of static work.

Nutritional requirements

The energy requirements of a forest worker are shown in Table 15. Carbohydrates are the best source of energy, giving a higher efficiency of muscles at combustion than fat. A person with a very high calorie consumption must replenish this by means of fat that supplies more than twice as many calories per gram as carbohydrates and proteins. In addition the body requires very essential minerals and vitamins.

Requirements for water, i.a. because of perspiration, are regulated by thirst. It is common at heavy work, particularly in warm weather, that the worker is not taking sufficient liquid for a replacement of the losses that have occurred because of perspiration and the regulation of body temperature. Dehydration reduces work capability.

Perspiration at forest work amounts to 0.1-0.4 l/h. In steel works and foundries perspiration is considerably higher.

Table 15. Calorie requirements per day in various occupations.

At rest (basic metabolism of body)	1500-2000 kcal
Sedentary work	2000-3000 "
Normal industrial work	3000-4000 "
Heavy industrial work	4000-5000 "
Manual forest work	5000-6000 "

Briefly on pulse rate at rest and at work

Under various circumstances pulse rates at rest normally vary betwen 50 and 80 per minute.

Under work conditions pulse rates increase in proportion to work load. This expresses itself in the oxygen intake and correponding energy requirements.

During felling and processing in the forests pulse rates vary largely between 100 and 160 per minute. Rates above 125 per min are often incompatible with lengthy work. Breaks or shifts to light work are then necessary. The average daily pulse rate of a feller is usually about 125 per min. Measurements of pulse rates are valuable information at investigations of the physiological work load, e.g. to clarify the work alternative that is preferable from a physiological point of view at studies of methods under heavy or hot working conditions.

References

Ager H:son B (1963) Preparation of timber landings on ice. Studia Forestalia Suecica No 1/1963. Skogshögskolan, Stockholm 1963

Ager H:son B, Nilsson N-E and von Segebaden G (1964) Description of some for logging operations important characteristics of forest stands, trees and terrain in Sweden. Studia Forestalia Suecica No 20/1964. Skogshögskolan, Stockholm 1964

Ager H:son B (1972) Tree and branch data relevant to logging operations. Report No 50. Department of Operational Efficiency, The Swedish University of Agricultural Sciences, Garpenberg 1972

Alriksson B-Å (1982) Mindre gallringsmaskin för fällning och sammanföring. SKOGEN, nr 10/1982, Stockholm

Alriksson B-Å (1983) Gör hästjobb i gallring. SKOGEN, nr 3/1983, Stockholm

Andersson, S (1982) Ny teknik i skogen, Sekretariatet för framtidsstuider Trosa, 1982

Anon Husqvarna skogsteknik, 1983

Anon Joint Committee (1965) Portable and semi-portable wood chipping machines. Part II. FAO/ECE/Log/161. Joint Committee on Forest Working Techniques and Training of Forest Workers, Geneva 1965

Anon FAO (1980) Chainsaws in tropical forest. FAO Training Series No 2, Rome 1980

Anon Skogsarbeten Resultat nr 17/1983, Stockholm

Anon Billeruds rapport Helträdsaktuellt. 1983

Arnelo N and Banner K (1967) Analys av maskinell upparbetning i mobila och semistationära anläggningar. Kungliga Domänstyrelsen, Arbets- och maskintekniska avdelningen. Stockholm 1967

Arvidsson A m fl (1983) Thinning techniques - Final report from a research project of the Nordic Forest Work Study Council 1978-82. The Swedish University of Agricultural Sciences, Department of Operational Efficiency. Report 156. Garpenberg 1983

Axelsson S-Å (1967) Analysis of Vibrations of Power-saws. Research Notes No 31/1967. Royal College of Forestry, Department of Operational Efficiency, Stockholm 1967

Axelsson S-Å (1971) Evaluation of Logging-machine prototypes. Arbomatik processer (LRA). Woodlands Report WR 35, Pulp and Paper Research Institute of Canada, Canada

Axelsson S-Å (1972) Repair Statistics and Performance of New Logging machines. Koehring short-wood Harvester. Logging Research Reports LRR 47, Pulp and Paper Research Institute of Canada, Canada

Bendz M and Yttermyr B (1966) Skogsarbetskraftens arbets- och servicefärder. Report No 30. Department of Operational Efficiency, Skogshögskolan, Stockholm 1966

Berlyn R W (1965) The effect of variations in the strength of the Bond between Bark and Wood on Mechanical barking. Research Note No 54, Pulp and Paper Research Institute of Canada. Woodland Research No 174, Montreal, Canada

Bjerkelund T C (1965) The tree-length system Operating features. Woodlands Review, May 1965, Quebec, Canada

Björklund E (1968) Skidding resistance for trees and stems. Research Note No 32/1968. Department of Operational Efficiency, Royal College of Forestry, Stockholm 1968

Bojanin S (1983) Arbeitsanalyse beim Transport von Blockholz mittels der Lastkraftswagen, Jugoslavien. XVII Internationale Symposium über Mechanisierung der Forstnutzung in Zalesina bei Delnice, Jugoslavien, 1983

Bol M M G R (1978) Simulation techniques Applied to Forestry Problems. Department of Forestry Technique and Forest Products, Agricultural University, Wageningen, the Netherlands, October 1978

Board of Technical Development, Sweden Proposal for a Forest Techniques Program 1984/85-1986/87, Info 363, 1983

Bredberg C J and Moberg L (1972) Felling heads designed for simultaneous handling of several trees. Royal College of Forestry, Department of Operational Efficiency. Research Notes No 51/1972, Stockholm

Callin G and Forslund S (1968) Kraftbehov vid kvistning med skärverktyg med rak och sned egglinje. Report No 34. Department of Operational Efficiency, Royal College of Forestry, Garpenberg 1968

Conway S (1982) Logging Practices. Principles of Timber Harvesting Systems. San Francisco, California

Corcoran Th J (1983) Harvest system and road design for a regulated wood flow. USA "XVII Internationale Symposium über Mechanisierung der Forstnutzung in Zalesina bei Delnice, Jugoslavien, 1983"

Cornides G (1973) Comprehensive Development. Planning of Forestry and Timber-Industry in Hungary. FAO/ECE/ILO Joint Committee, Geneva, Budapest 1973

Christov S (1983) Möglichkeiten zur Steigerung der Effektivität beim LKW Transport von Holz in VR Bulgarien. "XVII Internationale Symposium über Mechanisierung der Forstnutzung in Zalesina bei Delnice, Jugoslavien 1983

Croisé R (1972) Reflexions sur Les Routes Forestiéres. Centre Technique du Génie Rural des eaux et des Forést. Revue Forestiére française numéro special 1972

Dahlin B (1983) Översikt över metoder för kvistning av klenvirke. Review of delimbing methods for small trees. The Swedish University of Agricultural Sciences, Department of Operational Efficiency, Garpenberg 1983

Dehlén R, Herlitz A, Johansson I, Långström B and Regnander J (1982) Through-debarking of pulp wood. Debarking results and protective effect against bark beetles, as well as some aspects upon economics and ergonomics. Report No 143, The Swedish University of Agricultural Sciences, Department of Operational Efficiency, Garpenberg 1982

Delin B H (1966) Garpnäven visar vägen. Svensk trädklippare utformas i Garpenberg, SKOGEN nr 11/1966, Stockholm

Eisenhauer G (1981) Organisation, Durchführung und erste Ergebnisse einer Grossmodellversuchs zur Rationalisierung der Schwachholzernte. Institut für Arbeitswissenschaft (Iffa), Reinbekk. Tessaloniki 1981

Embertsén S (1976) Logging and Transport in the Kramfors Forests of SCA 1911-1965. Studia Forestalia Suecica nr 134/1976. Royal College of Forestry, Stockholm 1976

Equipment Denis inc. (1983) SJ-24 Saw felling head, Quebec, Canada

Eriksson L (1981) Strip roads and damages caused by machines when thinning strands. Report No 137. The Swedish University of Agricultural Sciences, Department of Operational Efficiency, Garpenberg 1981

Filipsson S (1983) Tillvaratagande av träd i röjningsbestånd. Studie av ett entreprenörsystem för flisning med manuellt matad flishugg. Stencilerna 203 och 249/1983. Sveriges lantbruksuniversitet, Inst f skogsteknik, Garpenberg

Focus Materia (1965) Atmosfären, Focus-Materien, pp 484-509, Stockholm 1961

Grammel R (1975) Aspects of multi-purpose logging machines in central Europe. Tim/EFC/WP1/SEM2/R24, Geneve. Symposium on Multi-purpose logging machines, Stockholm 1975

Grayson A J (1971) Labour Productivity Methods of measurement of labour productivity. Log/WP.1/16 17. March 1971, Geneve. Joint Committee meeting Geneve 1971

Haarla R (1973) The Effect of Terrain in the Output in Forest Transportation of Timber. Acta Forestalia Fennica Vol 128, 1973, Helsingfors, Finland

Hafner F, Mihǎc B (1968) Mehanzovani Transport Drveta, Jugoslovenski Poljoprivrednosumarski Centar Beograd, 1968

Hakkila P (1972) Progress Report on the Joint Scandinavian. Program for Marginal Wood Resource Utilization. The Finnish Forest Research Institute, Helsingfors 1972

Harricana Metal Inc (1983) Circular saw Feller-Buncher, Quebec, Canada

Harris G J (1965) Hydraulic barking symposium on mechanical barking of Timber. log Symp 2/12 Geneve, Helsingfors 1965

Hedbring O, Nilsson P O, Åkesson H (1968) Analys av några avverkningssystem för gallring. Logging Research Foundation, Redogörelse nr 4 1968

Hedegård B (1951) Motorers effekt och dragkraft - några grundläggande begrepp. Meddelande från SDA 44/1951, SST 1951 Stockholm

Hedman L (1983) Hästkörning i gallring under vinterförhållanden. "Småskogsnytt" nr 2/83. Stencil 227. Sveriges lantbruksuniversitet, Inst f skogsteknik, Garpenberg

Herpay I (1981) Tendenz der technischen Entwicklung in Mechanisierung der Forstnutzung in Ungarn. Das XV. Int. Symp. in Forsttechnik. Tessaloniki 1981

Hilf H H (1957) Arbeitswissenschaft. Grundlagen der Leistungsforschung und Arbeitsgestaltung. Carl Hanser Verlag, München 1957

Horncastel D C (1965) The full-tree system. Operating features. Woodlands Review Pulp and Paper Magazine of Canada, May 1965

Ionascu Gh., Antonoaie N, Ignea Gh (1982) Instalatii cu cablu, Pentur Transport de Lemn şi materiale Universitatea Din Brasov, Bucuresti 1982

Ievins J K et al (1976) Peculiarities of Harvesting technology in felling areas with a great number of small-sized trees. Riga 1976

Jonsson T (1929) Massatabeller för träduppskattning, Stockholm 1929

Kaldy J (1978) Munkave'delem az erdōgazdaságban Mezōgazdasagi Kiadő. Sopron 1978

Kaminski E (1981) Intergration der Waldwirtschaft und Holzindustrie. Universität für Landwirtschaft in Warszawa. Tessaloniki 1981

Kempe C (1967) Forces and damage involved in the hydraulic shearing of wood. Studia Forestalia Suecica No 55/1967, Skogshögskolan, Stockholm

Kilander K (1961) Time Consumption Variations for Felling of unbarked Timber in Northern Sweden. Meddelande nr 71, Forskningsstiftelsen Skogsarbeten, Stockholm 1961

Knigge W, Schulz H (1966) Grundriss der Forstbenutzung. Der Universität Göttingen. Hamburg und Berlin 1966.

Kolbas N S (1983) Effective Methoden der Bautechnik und der Nutzanwendung der Waldwege. USSR "XVII Internationale Symposium über Mechanisierung der Forstnutzung in Zalesina bei Delnice, Jugoslavia 1983"

Krivec A (1972) Mehanizirano Nakladanje Pri prevozu lesa. Biotehniska Fakulteta V Ljubljani, Ljubljana 1972

Kubasak E et al (1976) Technika a technológia výroby dreva. Výskumný ústav lesného hospodárstva, Zvolen 1976

Kubiak M (1976) Transport drewna wgospodarstiwie leśnym. Warszawa 1976

Larsson M et al (1983) Slutavverkning enligt träddelsmetoden. Resultat nr 14/1983. Forskningsstiftelsen Skogsarbeten (Logging Research Foundation) Stockholm 1983

Liss J-E, Risberg S (1983) Traevenderen - småskalig teknik för hantering av träd och stammar - metod och arbetsstudie. Dept of Operational Efficiency Stencil no 239/1983. Swedish University of Agricultural Sciences, Garpenberg 1983

Lundgren N (1964) Arbetsfysiologi och bioteknologi NKI, Malmö 1964

Luthman G, Olsson-Lokind P & Wesslén G (1942) Studier i skogsbrukets arbetslära. Undersökning vid Värmlands skogsarbetsstudier, Stockholm 1943

Mattsson-Mårn L (1956) Utveckling av några den skogliga arbetslärans grundläggande tankegångar. Materialsamling. Department of Science of Forest Labour. Royal School of Forestry, No 18, Stockholm

McMillin C H (1978) Complete Tree Utilization of Southern Pine. Proceeding of a Symposium. New Orleans, Louisiana 1978

Mihác B (1975) Application of processing machines for cutting, production and eventual transportation of wood. Tim/EFC/WP.1/SEM.2/R 19, Geneve. Symposium on Multi-purpose logging machines, Stockholm 1975

Moltesen P (1965) Chemical Barking. Log symposium 2/2 Kap IIA on mechanical barking of Timber, Joint Committee, Geneve, Helsingfors 1965

Muszunski Z (1976) Skladnice Drewna Skrypt Dea Studentow Leśnictwa wyższych szkol Molniczych, Krakow 1976

Myhrman D (1968) Report on testing the moving of trees vertically posited carried out in Garpenberg in October and November 1968. Forskningsstiftelsen Skogsarbeten, Stockholm

Newham R M (1972) Simulation studies in a logging program for the Canadian Forestry Service. Forestry Chronicle 48:26-29

Nilsson P O (1968) Analysis of possible logging systems for highly mechanized thinning in Sweden. Joint FAO/ECE/ILO Committee on Forest Working Techniques and Training of Forest Workers, 7. (Warsaw, 1968) Session, Geneva 1969

Nilsson P O, Hyppel A (1968) Studies on decay in scars of Norway spruce. Svenska Skogsvårdsförbundets Tidskrift 1966, Stockholm

NSR (1978) Forest Work Study nomenclature. Nordiska skogsarbetsstudiernas Råd (NSR), Helsinki, København, Stockholm, Oslo 1978

Pampel W (1978) Modellbetrachtungen zur Entwicklung der Langholzabfuhr in der DDR. XII Internationales Symposium. Mechanisierung der Forstnutzung, Tharandt Technische Uiversität. Dresden 1978

Pestal E (1961) Seilbahnen und Seilkrane für Holz und Materialtransport. Hochschule für Bodenkultur, Wien und München 1961

Platzer H B (1970) Der mechanisierte, zentrale Aufarbeitungsplatz für Schwachholz, Forstarchiv Nr 3/1970

Platzer H B, Wipperman H J (1972) Principle of centralized and mechanized conversion. Tim Symposium 2/31 9 March 1972, Geneve. Symposium on co-ordination between Forestry and the wood-using industries. Helsinki 1972.

Popa M V (1979) Informative Systems and Economical Acitivity Analysis of Forestry and Forestry Ezploiting Unities, University of Brasov, Bucharest 1979

Putkisto K (1958) Puttavaran Kuljetus Pyörätraktorilla. Metsätehon Oppaita, Helsinki 1958

Putkisto K (1970) Finsk drivningsteknik vid gallring och utveckling av denna XII Nordiska Skogskongressen, Helsinki 1970

Ronay E (1978) Klassifizierung der produktionstechnischen Bedingungen in der Slowakien und die Weiterentwicklung der Metoden der Holzernte, des Holztransportes und des Holzausfornung. Hochschule für Forst- und Holzwirtschaft, Zwolen 1958

Samset I (1970) Timber transportation with a Sikorsky S61N helicopter in mountainous regions of Norway. Report on Forest operations research No 9. The Norwegian Forest Research Institute, Vollebekk, Norway. Tidskrift for Skogbruk Nr 1, Oslo Norge 1970

Samset I (1981) Winch- and cable systems in Norwegian forestry. Reports of the Norwegian Forest Research Institute 37.1, Ås/NLH Norwegian 1981

Scholander J (1972) Compressive strength of some forest soil. Report 59/1972. Department of Operational Efficiency. The Swedish University of Agricultural Sciences, Garpenberg

Scholander J (1973) The bearing capacity of some forest soils for wheeled vehicles. Some technical aspects and consequences. Report No 64, Department of Operational Efficiency, Skogshögskolan, Stockholm 1973

Silversides C R (1965) Trends in wood. Woodlands Review, October 1965. Quebec, Canada

Silversides C R (1965) Chipbarkers, Symposium on mechanical barking of Timber, log Symposium 2/16 Geneve, Helsingfors 1965

Skogsarbeten, Forskningsstiftelsen (1969) Terrängtypschema för svenskt skogsbruk. Redogörelse nr 9/1969. Stockholm

Sprängare B, Troedsson H (1967) Röjning för framtida avverkningssystem. Skogen 5/1967. Stockholm

Staaf A (1953) En undersökning av sambandet mellan arbetsåtgång och behandlingsareal vid skogsvårdsåtgärder, Norrlands Skogsvårdsförbunds Tidskrift 1953:2, Stockholm

Staaf A (1960) Einige Gedanken zu forstlichen Betriebsplanung in Schweden. Forstarkiv 6. 1960, Hannover. Några synpunkter på skoglig driftsplanläggning, Skogen 1/1960, Stockholm

Staaf A (1962a) Låt lämpliga hjul rulla i skogen, Skogen 14:1962, Stockholm

Staaf A (1962b) Nytt traktorfordon för virkestransport, Skogen 24:1962, Stockholm

Staaf A (1963) Lastredskap för terrängtransport. Värmlands skogsarbetsstudier, Filipstad, 1963

Staaf A (1965a) Mekanisering av avverkningsarbetet i yngre bestånd. Installationsföreläsning i skogsteknik 9.4.1965, Hedemora 1965

Staaf A (1965 b) Samordning av mekaniserade arbetsoperationer vid avverkning och transport, Skogen nr 2/1965, Stockholm

Staaf A (1965 c) Barking Techniques: Technical Methods available or under development. Log Symposium 2/28 Geneve. Joint Committee on Forest Working Techniques and Training of Forest Workers, Symposium on Mechanical Barking of Timber. Helsingfors, Finland 1965

Staaf A (1969) Hjulet och skogsmarken. Jord och Skog. Traktorjournalen 1/1969, Linköping

Staaf A (1972) Drivning, avverkning och transport i skogsbruket. LT's förlag, Borås 1972

Staaf A (1979) Skydd och säkerhet i Skogsbruket. LT's förlag, Stockholm 1979

Staaf A (1982) Dörfer in der Mitte Schwedens. Eine soziologische Bewertung der Einflüsse der Forsttechnik auf die Dörfer in der gemeine Torp in den Jahren 1945-1980. XVI Internationales Symposium für Forstnutzung, Schönbrunn, Wien 1982. Inst f skogsteknik. Stencil nr 188/1982, Skogshögskolan, Garpenberg

Staaf A (1983 En skogsteknisk historik. Komplement till boken "Skogsbruk i omvandling". Institutionen för skogsteknik, stencil nr 251/1983. Skogshögskolan, Garpenberg

Steinlin H (1969) Beeinflussung des Gebrauchswertes von Faserholz zwischen der Fällung im Wald und dem Beginn des Fabrikationsprozesses. Thinning and mechanization, IUFRO-meeting Royal College of Forestry, Stockholm 1969

Stergiadiadis G C, Efthymiou P N, Katenidis K B (1981) Möglichkeiten zur rationellen Mechanisierung der Forstnutzung in Griechenland. Das XV Internationale Symposium "Mechanisierung bei der Forstnutzung". Aristoteles Universität, Tessalonike, 1981

STU (Styrelsen för teknisk utveckling) (1983) SKOGSTEKNIK Programförslag 1984/85-1986/87. Information 363-1983, Stockholm

Sundberg U (1952-1953) Studier i skogsbrukets transporter. SST 1952:4, 1953:1, Stockholm

Sundberg U (1957) The Mechanical barking of timber. Joint Committee on Forest Working, Techniques and Training of Forest Workers. FAO/ECE/Log/66, Geneve 1957

Sundberg U (1970) Interim Report on a Study of Full Mechanization of the First Thinning. No 43/1970, Department of Operational Efficiency, Skogshögskolan, Garpenberg

Sundberg U (1982) A Study on Cost of Machine Use in Forestry - Proposing fuel consumption as cost determinant. Department of Operational Efficiency Report No 142/1982, The Swedish University of Agricultural Sciences, Garpenberg 1982

Taraldrud H (1972) Mechanical limbing in thinning operations. (Rispekinteren) Report on Forest Operations. Research No 11 The Norwegian Forest Research Institute, Ås/NLH Norge. Tidskrift for skogbruk No 4, 1972, Oslo

Tekniska Nomenklaturcentralen (1969) Skogsordlista. Glossary of forest terms Swedish-English TNC 43, Lund 1969

Tomanič S (1974) Racionalizacija Rada. A Work Study of Felling, Primary Conversion and skidding of wood. University of Zagreb, 1974

Toyo Rinno (1983) Tree Pruning machine. Toyo Rinno Co Ltd., Fukuoka Japan 1983. Skogsarbeten (1983) Resultat nr 29/1983. Forskningsstiftelsen Skogsarbeten, Stockholm

Trzesniowski A (1981) Einsatz landwirtschaftlicher Traktoren bei Forstarbeit. Forstliche Bundesausbildungsstätte, Ossiach, Österreich. Tessaloniki 1981

Voronitsin K I, Vorobyev I V (1965) The Effect of the Season on Barking. Institute for Mechanization of the Timber Industry (TSNIIME) Moscow. Symposium on mechanical barking of Timber. Helsingfors, Finland 1965 Joint Committee on Forest working Techniques and training of Forest Workers. Log Symp 2/19, Geneve

Wassilev W (1981) Analyse der Kräfte, die auf den Tragzugseil der leichten Kurzstreckenseilanlagen einwirken, bei Stamrückung mit Anheben der Vorderende. Institut für Forstwirtschaften, Sofia, Tessaloniki 1981

Wibstad K (1983) Transportkosten der norwegischen Waldindustrie. XVII Internationale Symposium über Mechanisierung der Forstnutzung in Zalesina bei Delnice, Jugoslavien 1983

Wiklund M (1967) Power and damage in felling and barking with knife and shear-type equipment. Forskningsstiftelsen Skogsarbeten, Redogörelse nr 9/1967. Stockholm

Wiksten N Å (1977) Using wood for energy. Nova Scotia Department of Lands and Forests, Nova Scotia, Canada Cat/77/141/400

Åstrand I (1960) Aerobic Work Capacity in Men and Women with Special Reference to Age. Acta Physiol. Scand. 49, 1960

Östbergs Fabriks AB (1982) ÖSA-System nr 1/1982 Information från Östbergs Fabriks AB, Alfta Bollnäs 1982

Österlöf P (1981) Träddelsgallring med gripsåg på långkran. Forskningsstiftelsen Skogsarbeten